The Empathic Screen

The Empathic Screen

Cinema and Neuroscience

By
VITTORIO GALLESE AND MICHELE GUERRA

Translated by
FRANCES ANDERSON

OXFORD
UNIVERSITY PRESS

Great Clarendon Street, Oxford, OX2 6DP,
United Kingdom

Oxford University Press is a department of the University of Oxford.
It furthers the University's objective of excellence in research, scholarship,
and education by publishing worldwide. Oxford is a registered trade mark of
Oxford University Press in the UK and in certain other countries

© 2015, Raffaello Cortina Editore
Originally published as *Lo Schermo Empatico* di Vittorio Gallese e Michele Guerra,
Raffaello Cortina Editore, 2015.

© Oxford University Press 2020

The moral rights of the authors have been asserted

First Edition Published in 2020

All rights reserved. No part of this publication may be reproduced, stored in
a retrieval system, or transmitted, in any form or by any means, without the
prior permission in writing of Oxford University Press, or as expressly permitted
by law, by licence or under terms agreed with the appropriate reprographics
rights organization. Enquiries concerning reproduction outside the scope of the
above should be sent to the Rights Department, Oxford University Press, at the
address above

You must not circulate this work in any other form
and you must impose this same condition on any acquirer

Published in the United States of America by Oxford University Press
198 Madison Avenue, New York, NY 10016, United States of America

British Library Cataloguing in Publication Data

Data available

Library of Congress Control Number: 2019935965

ISBN 978–0–19–879353–3

Links to third party websites are provided by Oxford in good faith and
for information only. Oxford disclaims any responsibility for the materials
contained in any third party website referenced in this work.

The translation of this work has been funded by SEPS
SEGRETARIATO EUROPEO PER LE PUBBLICAZIONI SCIENTIFICHE

Via Val d'Aposa 7 –40123 Bologna –Italy
seps@seps.it – www.seps.it

To Alessandra, Clara, Giovanni, Lea, Leonardo, and Lorenzo

Foreword
The Neuroscientific Turn

Consilience—the integration of knowledge generated across disciplines—has been the talk of the town in some quarters for some time now. But rarely has it been executed with such seriousness and panache as it has been in *The Empathic Screen*, a collaboration between neuroscientist Vittorio Gallese and film scholar Michele Guerra. Here they bring into dialogue Gallese's command of contemporary neuroscience, especially the line of research emerging from the discovery of mirror neurons— in which Gallese played a central role—with Guerra's expert knowledge of the long tradition of film theory, stretching back to the origins of cinema itself around the turn of the twentieth century. And, far from being mere promissory chit-chat, the dialogue that Gallese and Guerra conduct—between themselves, their immediate collaborators, the wider research community, and the various historical traditions of film theory—is methodologically meaty. By combining close analysis of film style with theoretical reflection and neuroscientific experimentation, Gallese and Guerra walk the consilient walk as well as talk the interdisciplinary talk. They seek to triangulate cinematic experience through the integration of humanistic and neuroscientific methods. But their dialogue is not closed or exclusive: through the clarity of the book's writing and the glossary of specialist terms provided by the authors, readers not expert in either or both neuroscience and film theory are welcomed into the conversation.

If the concern with interdisciplinarity and consilience is a hallmark of contemporary academia, in other ways *The Empathic Screen* roots itself in a wide-ranging and deep engagement with film history, from its infancy to the most current developments. An abiding concern of film theory has been with the *specificity* of the medium of film—the feature or features of it that make it distinctive and not quite like any pre-existing vehicle of representation or medium of art. This fascination with specificity is evident even in the work of those theorists who have been officially skeptical about the idea of medium specificity, like Noël Carroll, and even in those specific trends in film theory where it has not appeared to be primary, such as contemporary post-structural film theory.

Almost everyone drawn to thinking about cinema ends up facing the question: what makes movies special? What is the distinctive power of movies?

Gallese and Guerra approach this question by posing two questions of their own: *why* do movies fascinate us? And *how* do they engage us? Their answer to both questions draws on the notion of *embodied simulation*, a theory of human perception of and action in the world, including our interaction with other human agents. According to the theory of embodied simulation, one of the most basic ways in which a human agent relates to other agents is by simulating the movements made, sensations felt, and emotions experienced by them. And far from being a cool, abstract, theoretical affair, when humans simulate they do so in embodied fashion, their vision, audition, and other senses working together, allowing them to grasp shapes and textures, feel the affective states of others, and kinesthetically mimic their conspecifics. Embodied simulation is the foundation of human intersubjectivity in general, and empathy in particular—long an object of fascination for film and literary scholars—and it operates in multisensory fashion. As Gallese and Guerra put it, the neural integration of our sensory modalities is the rule rather than exception.

In the first instance, embodied simulation is a theory of human perception, action, and interaction in general, and part of Gallese and Guerra's agenda is to show how, to a large degree, films are designed to allow the seamless carryover of the mechanisms of embodied simulation from the world itself to the motion picture world. Before we can speak, we can perceive and move and resonate to the movements of others through our mirror mechanisms, and cinema holds us at this primal level. But where then is cinematic specificity to be found? It is to be located, Gallese and Guerra argue, in the form of *liberated* embodied simulation. Moving pictures allow us to transcend the simulation scenarios available to us through our concrete encounters with real others in proximate situations, affording us the opportunity to simulate the experiences of real others in situations remote from us in space and/or time (in documentary films), and of imaginary others in imaginary places (in fiction films). In this way our empathic horizons are expanded. But it is not only in terms of make-belief and remote content that films can liberate our embodied simulative capacity. Films may also expand and enrich our simulative repertoire through technical and stylistic innovation within filmmaking, as Gallese and Guerra discuss in relation to the Steadicam and action cameras, as well as technological innovations at once influenced by and impinging upon cinema, as in the recent rise of mobile screen technologies.

In their exploration of mirror neurons, and of the embodied nature of biological cognition, Gallese and Guerra bring to bear two great contemporary discoveries on our understanding of cinema. They begin and end their study by reflecting on cinema in the light of one of the frontiers of contemporary archaeological discovery: the modern investigation of cave paintings—their whereabouts, age, character, and functions—a story beginning just a few decades before the Lumière brothers first public screening, and continuing to unfold across the twentieth and twenty-first centuries. The extraordinary images found in Altamira, Lascaux, Chauvet, and many less well-known locations remind us that behind the modernity of cinema—its dependence on technology that was not invented until the nineteenth century—lie much more ancient human impulses: impulses to depict, to imagine, to narrate, to enrapture. This combination of ancient and modern finds an echo in other aspects of Gallese and Guerra's work: contemporary neuroimaging meets the evolved biotechnology of the brain on one timescale, and the archaeology of cinema history on another. In this way, the example of cave art functions as both a model for research on cinema in its most ambitious form, as well as forming part of the story of cinema in its broadest scope.

Gallese and Guerra's project is pluralist—one might even say ecumenical—in character, making reference to a wide range of authors across the history of film theory, and insisting upon the importance of 'profitably conjugat[ing] the experiential dimension' of cinema with the 'study of the underlying subpersonal processes and mechanisms expressed by the brain and its neurons'. But there is no doubting the book's particular relevance for the tradition of cognitive film theory. As that label attests, when cognitive film theory emerged in the mid-1980s, it bore the imprint of the first wave of cognitive science, with its emphasis on computation and the human mind as an information processor. Broadly speaking, cognitive film theory has tracked developments in its parent domain, by subsequently registering the importance of emotion, embodiment, and evolution to cognition. The rise of cognitive *neuro*science is a key part of this developing picture. *The Empathic Screen* is not the very first work of 'neurocinematics', but it is among the very first book-length studies, distilling the results of one of the earliest sustained efforts to apply neuroscientific methods to the study of film, and to integrate those methods with humanistic ones. Gallese and Guerra thereby help to forge a third culture, bridging the 'two cultures' of the sciences and the humanities. They emphasize that this project in the larger sense has barely begun. But as a first step, it's a pretty impressive one.

Murray Smith

Acknowledgments

This book was born from over five years of research and almost daily discussions on the relationship between art and neuroscience. As we say in the Introduction, it is a milestone on our road, not the final destination. Over the years we have had numerous opportunities to present our thoughts and research at various conferences and seminars in Italy and abroad, and we have published papers on periodicals of many disciplines, including cinema, neuroscience, psychology, and cognitive science. Many of the concepts contained in this book came to light as a result of these activities and encounters with friends, colleagues, and students. For this, we would like thank Hava Aldouby, Luigi Allegri, Marianna Ambrosecchia, Massimo Ammaniti, Monica Angelini, Martina Ardizzi, Paul Auster, Nurith Aviv, Karin Badt, Paolo Benvenuti, Cristina Berchio, Bernardo Bertolucci, Colin Blakemore, David Bordwell, Dominique Budor, Marta Calbi, Roberto Campari, Enrico Carocci, Francesco Casetti, Maarten Coëgnarts, Hannah Chapelle Wojciehowski, Michele Cometa, Valentina Cuccio, Elena Dagrada, Antonio Damasio, Adriano D'Aloia, Roberto De Gaetano, Marta Dell'Angelo, Ophelia Deroy, Giuseppe di Cesare, Cinzia Di Dio, Sjoerd Ebisch, Ruggero Eugeni, Francesca Ferri, Joerg Fingerhut, David Freedberg, Chris and Uta Frith, Shaun Gallagher, Leonardo Gandini, Alexander Gerner, Ori Gersht, Gabriella Gilli, Art Glenberg, Alvin Goldman, Patrick Haggard, Mark Hansen, Anne Harrington, Katrin Heimann, Siri Hustvedt, Dan Hutto, Noah Hutton, Marc Jeannerod, Stephen Kosslyn, Peter Kravanja, George Lakoff, Ludovica Lumer, Harry Malgrave, Anthony Marcel, Antonella Marchetti, Davide Massaro, Thomas Metzinger, Jonathan Miller, Akira Murata, Ugo Morelli, John Onians, Peppino Ortoleva, Juhani Pallasmaa, Jaak Panksepp, Francesco Parisi, Michael Pauen, Antonio Pennisi, Roberto Perpignani, Guglielmo Pescatore, Héctor Perez Lopez, Andrea Pinotti, Carl Plantinga, Vassilis Raos, Vasudevi Reddy, Giacomo Rizzolatti, Magali Rochat, Jelena Rosic, Beatrice Sbriscia Fioretti, Mariateresa Sestito, Corrado Sinigaglia, Patrícia Silveirinha Castello Branco, Barry Smith, Murray Smith, Tim Smith, Mark Solms, Daniel Stern, Anna Strasser, Pia Tikka, Colwyn Trevarthen, Manos Tsakiris, Sebo Uithol, Alessandra Umiltà, Joe Walker, Sigrid Weigel, Dan Zahavi, and Semir Zeki.

We are grateful to Frances Anderson and to the Oxford University Press staff for working so carefully on this book: Martin Baum, Charlotte Holloway, Karen Moore, and our copyeditor Jayne MacArthur.

The electroencephalogram (EEG) experiments described in Chapters 3 and 4 were financed by a European Commission Marie Curie grant, "TESIS: Toward an Embodied Science of Intersubjectivity," and by the Chiesi Foundation of Parma. Should we have forgotten anyone, please accept our apologies—the oversight is entirely involuntary.

Table of Contents

Introduction   xv

1. Embodied Simulation: A New Model of Perception   1
2. Stilted Movements and Improbable Stares   53
3. Camera Movements and Motor Cognition   85
4. Cut and Harmony   117
5. Face and Hands   145
6. New Mediation, New Films, New Experiments   181

Glossary   203
References   209
Index   227

Introduction

Cave of Forgotten Dreams, directed in 2010 by Werner Herzog, is an extraordinary 3D film of the Chauvet-Pont-d'Arc cave paintings in Southern France. He brought state-of-the-art technology to the graffiti traced on the grotto walls around 30,000 years ago at the height of the Upper Palaeolithic period, breathing new life into those ancient images by recreating them in 3D. The film becomes increasingly intriguing as the exploring camera brings into focus the artists' intention, an attempt to communicate a sense of movement. There are approximately 100 drawings in the Chauvet cave, representing 13 species of animals, including horses, lions, bears, and panthers. Many of these drawings are not simply static reproductions of an anatomical form; there are scenes that communicate dynamism, action, and purpose. The series of images depicting lions hunting a herd of bison in flight capture the bodily dynamics of both and tell a story—a story that could well be an allegory of the human condition.

In 2012 Marc Azéma and Florent Rivère published an article on *Antiquity*, in which they suggested that the Chauvet cave drawings and other similar Palaeolithic paintings could well be considered the forerunners of cinema.[1] These prehistoric "filmmakers" used two different techniques to communicate a sense of movement; in one the drawings are superimposed and in the other they are sketched in series, generating a dynamic fluidity that anticipated the era of movie-making by millennia; and not only did these prehistoric artists anticipate cinema, they were also eons in advance of Marcel Duchamp or the Futurists.

This hypothesis is corroborated by the interpretation that the authors attribute to a prehistoric bone disk found in a cave in the Dordogne in 1868. This disk has a hole in the center and a chamois depicted on both faces: on one, the animal is drawn standing and on the other, lying down. In 2007, Florent Rivère conducted a simple experiment; he made a copy of this disk, passed a thread through the central hole, and then twirled the disk rapidly over and over. When he drew the ends of the thread tight, the disk rotated on its lateral axis, producing a form of animation in which the chamois seemed to lie down and

---

[1] Azéma and Rivère (2012). Other references to Chauvet, regarding questions dealt with in this book, can be found in Onians (2007).

jump up in rapid succession until the thread unwound completely. Azéma and Rivère interpreted this phenomenon as a possible forerunner of the thaumatrope (from the Greek thauma = wonder and trope = turner), an instrument that was invented in 1824 and which, together with other precinematographic devices, testifies to the human obsession with animating images, which rose to a peak in the 1800s.

It could be said then that a common thread, the natural propensity of man to use inert material to reproduce the living world through images, links these Palaeolithic artists to the Lumière brothers. This transformation is achieved by breathing movement into the images, or at least by suggesting that there is movement, as sculptors have done throughout the ages. Technological evolution could even be interpreted as a constant attempt to animate the material world through movement.

The authors of this book, by profession a neuroscientist and a film theorist, share a fascination with images that is primordial and atavistic, as Herzog shows so brilliantly in his film. The power they have, the attraction they exert, their ability to replicate or simulate experiences, movement, or indeed, the impression of movement they transmit, are themes that since the last century have found a form of radicalization in cinema from the aesthetic, cultural, social, and political points of view. It goes without saying that these aspects profoundly affect how we relate to the world and lead to a pressing question: What form does our relationship—our "contact"—take with these images that look so real, are close to hand, and obviously relate to our reality but still have the power to transport us into a purely virtual dimension? In this book we investigate this power by examining styles and techniques used by filmmakers with moving images from the earliest days to the present. The key question that we have posed is what type of intersubjective relationship exists between spectators and the worlds of make-believe they enter and exit with increasing frequency; cinema may not represent the first "machine" that created this kind of relationship, but it is certainly the one that has had the greatest impact on the senses of millions of people and on the collective imagination. The study of cinematographic intersubjectivity requires a search for the origin of the concept of intersubjectivity itself and an understanding of how human beings relate to the space in which they live, to the people and objects that surround them (and how they can do this just as well when watching a movie as in real life). Today new findings and innovations in the field of cognitive neuroscience can inject renewed vitality into these topics, which have been widely discussed in debates, in critical analyses, and in studies on the philosophy of cinema.

In fact, neuroscience has revealed that even at the sub-personal level of description, the level of neurons and cerebral circuits, human intelligence is closely linked to corporeality. Our corporeality is not merely a physical object endowed with extension, it requires experience to be complete. The human body is a living entity that acts in, and accumulates experiences of, an external world, and concepts such as "being," "feeling," "acting," and "knowing" describe diverse modalities of our relations with that world. This corporeality plays a decisive role in the simulations of which we are capable not only in our daily life, but also in our aesthetic and mediated experiences. In recent years the relationship between neuroscience and humanistic studies has given rise to multidisciplinary research based on a reassessment of the communication of art forms originating from embodiment; in other words, starting from forms of engagement of our brain–body system.

The involvement of neuroscience in art and aesthetics is a fairly recent development; research in this area has concentrated mainly on how the brain processes images. Semir Zeki is a pioneer in this field and to him goes the merit of a new line of research, "neuro-aesthetics."[2] This approach immediately became widely diversified; some neuroscientists employed art forms just as stimuli to gain a better understanding of how the brain functions, using paintings and sequences taken from movies, with no direct reference to their aesthetic and cultural qualities, to improve comprehension of the neurobiological basis of cognitive faculties not specific to art. Others, including Zeki himself, used brain imaging techniques such as functional magnetic resonance imaging (fMRI) to study the concepts of "beauty" and "aesthetic pleasure," and also, on a more general level, to study the neural mechanisms responsible for the visual perceptive analysis of various formal characteristics of works of art.

This book proposes an approach to cinema based on a different application of neuroscientific research methods. We were always convinced, from our earliest discussions on, that it was essential nothing should be lost of our exchanges and debates; nothing regarding the neuroscientific vocation of understanding how the brain–body system works nor the cinematographic vocation of studying significant aspects of the movie experience. With this premise, it was clear that we could not limit our analysis to a simple experiment, however well designed and elegant; it needed to be structured as a comparison between research methodologies and study traditions in order to create a method that would be both critical and productive.

---

[2] Zeki (1999). See also Zeki (2009).

We have defined our approach as *experimental aesthetics*, in which the notion of aesthetics is interpreted in accordance with its etymology: *Aisthesis*, the multimodal perception of the world through a combination of the senses.[3] This does not mean that neuroscience is in favor of discarding the great humanistic tradition for those reductionist theories that convert the contents of our aesthetic experiences into cerebral maps. Such things are fine for media headlines, as those which in recent years have contributed to creating a craze known as "neuromania" that predictably produced an equally violent wave of "neurophobia" in certain scholarly circles.[4] Participating in seminars and workshops, and talking to friends and colleagues over the years, we have become increasingly conscious of an aspect that may appear banal, but that unfortunately in all its disarming simplicity has eluded comprehension: The questions posed by scholars and researchers, whether they be neuroscientists, film historians, experts in mass media, architects, or physicists, refer to the countless ways in which humans interact with their external environment and are formulated according to their specific area of study or research. However, no single field of knowledge, no matter how extensive and exhaustive it may be, can claim to have "the" answer (always supposing that an answer actually exists), so the discussions that are now cropping up with increasing frequency between disciplines that appear to have little in common can do much to further research, study, and debate. As will become evident in this book, it would be hard to find a single theory of cinema or philosophical line of thought that has not stimulated our reflections and facilitated our approach, or served as a catalyst in unravelling certain problems, correcting the perspective on certain suggestions, not to mention helping us to understand fully the potential of others.

Instead of asking "What is cinema?", as did the great critic and theoretician André Bazin,[5] we opted to start with the question "Why do we go to the movies?", raised as far back as 1915 by the Harvard psychologist Hugo Münsterberg, a pioneer in the study of the cinema–brain relationship, who dealt with this subject in a short article that appeared one year before his major work, *The Film. A Psychological Study*, considered by many to be the first book on the theory of cinema.[6] As we contemplated this basic question, many others came to mind, such as "What does watching a movie really mean?", "To what extent and degree does the perceived artificiality of cinema cause us to deviate from the modalities we use to engage with our daily reality?", "Why are our

---

[3] Gallese (2015); Gallese and Guerra (2014).
[4] See Legrenzi and Umiltà (2009), and Aglioti and Berlucchi (2013).
[5] Bazin (1958–1962).
[6] See Jullier (2002, p. 7 and following).

emotions often affected more by what we see in a movie than by the same experiences in real life?", "To what extent is our viewing of a movie influenced by our personal identity, our background and experiences, our cultural conditioning?", "How relevant is the level of empathic and corporeal involvement to understanding how the technology of cinema works?", "What is the best way to study the involvement with the images and sounds of a movie and the subsequent immersion in the plot?"

At this point we asked ourselves whether neuroscientific research methodologies and instruments would be able to contribute to enriching and further developing our understanding of the mechanisms involved in viewing movies, if it is feasible to consider the corporeal involvement of the spectator as a measure of the salience and efficacy of a scene, and what strategies are used to engage the audience so that they empathize with the characters and the events in which they are involved.

This was the starting point for our theoretical considerations, our hypotheses, and a scientific research program which is still ongoing, the results of which we will discuss throughout the book. We will propose a robust hypothesis using a new perception model known as *embodied simulation*[7] to show how neuroscience can help us to better understand what cinema is and how its relationship with the audience works. This model can be applied both to our experience of the real world and to the fictional world of cinema; in fact, we believe it constitutes a basic functioning mechanism of the brain–body system in primates, including humans. By activating sensorimotor and visceromotor maps in the brain of the observer, this mechanism can facilitate the construction of a direct and non-linguistic relationship with space, objects, and the actions, emotions, and sensations of others; according to our hypothesis it is strongly involved in generating human imaginative abilities. Our theory therefore focuses on the starting point of the relationship with images, connected to the corporeal base of every type of relationship that can potentially be formed with the world's many aspects. We will present a study of film, starting from the sensorimotor and affective implications, identifying a series of neurophysiological mechanisms as a tool for the deconstruction and comprehension of the conceptual instruments usually adopted in debates on the various aspects of cinema. Embodied simulation, which according to neurophysiological evidence is the basis for an integrated and empirically proven model of our relationship with images and with film, is the foundation for most of the reflections

---

[7] See Gallese (2003, 2005, 2011), and Gallese and Sinigaglia (2011a).

in this book. Indeed, we suggest that it can clarify important aspects of the aesthetics of film: How it is constructed, how it is received.

In synthesis, the book revolves around five cardinal points:

1. Neuroscience can offer a genuine contribution to the redefinition and comprehension on new empirical grounds of fundamental and paradigmatic themes of the human condition, such as the perception of images and the construction of our network of relations with objects and other human beings. Action, perception, and cognition are terms and concepts describing modalities that are different but permanently bound by the embodied and relational essence of all living beings, including humans.
2. To be applied successfully, the neuroscientific approach to cinema proposed in this book must be critical, its potential and its heuristic limitations must be clear, and it must be open to dialogue and constructive collaboration with other disciplines such as the history of cinema and the theory of film, philosophy, and other humanistic traditions of study. The final objective must be to discover how to profitably conjugate the experiential dimension in the first person through a study of the underlying sub-personal processes and mechanisms expressed by the brain and its neurons, and promote investigations that can lead to progress in both film theory and philosophy.
3. This bottom-up approach has demonstrated that not only *looking* at the world, but also simply *seeing* objects and other people requires processes that are much more complex than the mere activation of the brain's visual fields. In other words, seeing the world means looking at the world with the intention of understanding it. Our visual experience of the world is the outcome of multimodal integration processes, in which the motor system is one of the key players.
4. The multimodal integration of what we perceive is triggered by the potentiality for action that we express corporeally, and therefore intentionally, in this world populated by other human beings similar to ourselves. We build non-verbal representations of our surrounding space, we have similar non-verbal relationships with objects and other beings like ourselves, using the functional base mechanism of embodied simulation that describes, from a functional standpoint, the neural mechanisms that ensure the clarity of our connections with the world around us, forming a dialectical relationship between the body and the mind, between subject and object, between you and me. It underlies important aspects of

empathy but cannot be identified as such as it has wider and more diversified areas of application.
5. Liberated embodied simulation, a particular form of embodied simulation, can help us to better understand the particularity and aesthetic insularity of our film experience, and shows where it deviates from and where it resembles the experience of what we call the "real world."

These points will be expanded upon in the main chapters of the book, starting with case studies from the history of cinema on which we have tested some of our ideas. We will revisit masterpieces past and present, each of which conveys a strong sense of cinema and the world, particularly through the processes with which they form spectators' experiences and how they become involved in the movie. This journey through the world of movies, from Buster Keaton to Alfred Hitchcock, from Delmer Daves to Robert Siodmak, from Stanley Kubrick to Jonathan Demme, from Ingmar Bergman to Jean-Luc Godard to Jan Švankmajer and the new forms of post-cinema and media, will help us to further investigate and question the forms of our presence at the cinema.

The book is divided into six chapters, followed by a short glossary of the technical neuroscientific and cinematographic terms used.[8]

The first chapter provides a review of the epistemological and neuroscientific notions needed to fully understand the following chapters. It also illustrates the theory of embodied simulation, which is crucial to comprehending our approach to cinema, movies and our experience of them. We have kept the technical language to a minimum while giving a concise summary of over 30 years of research that has culminated in a new concept of vision and intersubjectivity, providing the reader with the instruments to follow the propositions in the following chapters.

The second chapter examines the forms of subjectivity deployed by cinema and illustrates the methods and techniques used to achieve a credible overlapping of what the camera sees and what the spectator sees, delegating the camera to simulate the human body within the space of a shot.

The third chapter focuses on the various camera movements obtained with different techniques and technologies, and the type of motor resonance that these movements induce in the spectator. This chapter includes the results of two experiments[9] conducted using high-density electroencephalography

---

[8] Throughout the book gender-neutral masculine pronouns are used for simplicity of reading.
[9] Heimann et al., 2014, 2019.

(EEG) to study these forms of resonance evoked by camera movements produced with a dolly, a zoom, and a Steadicam.

The fourth chapter deals with editing techniques, another essential aspect of filmmaking. Once again, the quality of the spectator's motor engagement is studied in a state of continuity (represented by "continuity editing" of which the spectator should not be aware) and when viewing "jump cuts," one of the best-known deviations from the classic style. The chapter includes a discussion of the results of a high-density EEG research study conducted specifically on jump cuts.

The fifth chapter is dedicated to tactility and participates in the debate on *haptic visuality* with new elements that have emerged in recent years from neuroscientific research. The chapter examines close-ups and the texture of the filmed image in art films, mainstream animation, and experimental films.

The sixth and final chapter opens a window onto the future and offers a starting point for new theoretical reflections and experimental projects. A number of Hollywood's most influential filmmakers and producers are surmising that movies will undergo a revolutionary change with the advent of portable devices, the opportunities offered by technologies such as smartphones and action cams that have made their debut in the world of filmmaking, not to mention other devices and ambiences that will force filmmakers to rethink the forms of involvement at their disposal. In short, to cope with the technologies of today and tomorrow, cinema will need all the ability and flexibility for dialogue that it has always shown in the past; indeed, new technologies in this field owe much to its aesthetic models and narrative modules.

In short, *The Empathic Screen* is both a finishing point for certain reflections and a starting point for future developments; the theories and hypotheses discussed here are a catalyst for a series of new questions—including the role of sounds and music, which we did not address in this book—that we hope to tackle in the near future using neuroscientific research methodologies. One of the first questions that comes to mind concerns the role that social context plays in viewing images and movies in particular: For example, it would be interesting to study whether viewing the same movie in a group, as opposed to in an individual context, has a different effect on the embodied simulation mechanisms and interoception at the cerebral level. Then there is the correlated theme of mediation discussed in the last chapter in relation to the emergence of portable devices and the evolution of the current concept of screen. The neurophysiological and corporeal parameters could be measured during the viewing of both narrative and non-narrative films on devices with differing levels of interactivity and proxemics vis-à-vis their relation with the spectator.

Suspense is another topic that merits particular attention; it is a recurring theme in this book, particularly in the second and fourth chapters. The same scene, seen time and time again, will evoke identical feelings of anxiety and fear at each viewing, however familiar the spectator may be with the outcome. Suspense would appear to be based on a conflict between the cognitive and the corporeal dimensions of our relation to the images; studying this phenomenon empirically could reveal the underlying sub-personal mechanisms.

Finally, in the fifth chapter we suggest that enlarging an image can magnify the qualities of the materials and the plot, thus strengthening the empathic resonance and tactile projection mechanisms. A close-up of an image not only touches us more, both metaphorically and physically, but it also increases our desire to reach out and actually touch the image. The interactive viewing style of digital devices is modifying and possibly even enhancing these aspects; we are convinced that neuroscience will have an important role to play in supporting our understanding of the impact and implications of these new configurations.

Indeed, the more we worked on this book, the more we became aware of just how much still remains to be done.

1
Embodied Simulation
A New Model of Perception

Cinema, brain, and empathy

The power of cinema is such that we empathize not only with the experiences of other human beings, but even with those of animals. Take the case of Robert Bresson's *Au hazard Balthazar* (1966), which tells the story of the wretched life of a donkey named Balthazar and, in parallel, of that of Marie, his first and only friend. Their lives intertwine in a gradual descent into a world of violence, oppression, and humiliation. In minimalistic style, Bresson draws us into the situations in which Balthazar and Marie find themselves, situations that call for compassion, often focusing on details such as hands which touch, caress, and shake bodies and manes (Fig. 1.1), so that we see Balthazar's world also through his eyes and experiences.

This is one of the most fascinating aspects of cinema, and what transforms it into an art of the humanities. By telling a story through images in movement, not only can the filmmaker involve us in the story, he can even change our perception of the world. After watching Bresson's movie, the way we think of animals, the thoughts and sentiments we associate with them, how we relate to a real flesh-and-blood donkey, our potential relations with the animal world, will all probably be very different compared to how we felt before we saw the movie. But how is this possible? Where does this involvement with the characters that we see moving on the screen come from, this involvement that is not merely emotional, and what is it in this involvement that binds us to them?

Our thesis is that embodied simulation can make an important contribution to a theory of intersubjectivity and hence also to a theory regarding how films are received by the spectator. Embodied simulation is a basic functional mechanism of the brain by means of which part of the neural resources that are normally employed to interact with the world around us, shaping our relationships and relations, are reused for perception and imagination. Our understanding of the meaning of much of the behavior and the experiences of other beings relies on this reuse of the neuronal circuits on which our personal agentive,

Fig. 1.1 *Au hazard Balthazar* (Robert Bresson, 1966).

emotional, and sensory experiences are based.[1] We reuse our mental states and processes, represented in corporeal form, to attribute them functionally to others. Embodied simulation provides an integrated and neurobiologically credible framework for this type of intersubjective phenomena.[2]

The use of this model as a key to understanding and interpreting our receptivity of film is justified by the widely shared conviction that our approach to real life and to movies is based on perceptive mechanisms and on the underlying neurophysiological mechanisms that are very similar. The theory of embodied simulation lays the foundation empirically for a new model of perception that can be applied to many aspects of the way in which film works and how it is received by spectators. Another point in support of our hypothesis is that it can also be applied to forms of receptivity that are not necessarily just empathy.

According to Michael Haneke, Bresson's artistic work qualifies as specifically filmic due to the search for truth that he conducts through the systematic ambiguity of the situations in the plots and the way in which they are narrated in images.[3] In Haneke's view, the spectator's role can no longer simply be empathetic and receptive, as was the case in classic cinema and still is in the greater part of mainstream cinema. Bresson's narrative delivered by images (and also Haneke's, though in a different way) is, in fact, under-determined, ambiguous, and open but still able to guarantee a perfect adherence of form and content. And this, according to Haneke, is what makes the images so real. Spectators have to struggle with the narrative, and not just metaphorically, delving into

---

[1] For a discussion of the concept of neuronal exploitation and reuse, see Gallese (2008, 2014) and Anderson (2010).
[2] Gallese (2003, 2005, 2011); Gallese and Sinigaglia (2011).
[3] Haneke (2010, pp. 565–74).

their memory and their own ways of relating to objects and other people because, just as in real life, they don't know how the story is going to end.

It is probable that this projective relation modality is not restricted to the understanding of the narrative reality in the world of illusion. Owing to the involvement of our competence and sensibility as individuals in the field of relations, or in other words, thanks to our unique historic personal identity, this particular relation modality contributes both to building the sense that we attribute to the world and to the processing of our mental objects. Embodied simulation can make recordings in our implicit memories just as historical contingencies modify the style and rhythm of our relations with the world. In this respect, the embodied simulation that kicks into action when we observe others doing something, when we consider the consequences of those actions or think about doing, seeing or feeling something ourselves, updates our personal identity in the dynamic relation modality. It is both the line and the hand that draws it. Our hypothesis can therefore provide a coherent and neurobiologically established base for the diverse modalities with which we relate to what we see on the cinema screen.

The discovery of mirror neurons in the brain of the macaque,[4] followed by that of mirroring mechanisms in the human brain (of which more will be said later),[5] has shown that there is a neurobiological foundation for a direct modality of access to the meaning of the behavior and experiences of others. This direct access is independent of the explicit attribution of propositional attitudes such as desires, beliefs, and intentions that are typical of classic cognitivism's standard notion of intersubjectivity. Intersubjectivity is not completely covered in the linguistic-inferential mechanisms of the so-called "Theory of Mind."

Furthermore, the discovery of mirror neurons has made it possible to derive subjectivity from intersubjectivity at the sub-personal level of description, i.e., at a level that does not apply to the individual as a whole but just to the brain–body system. In other words, the mirror mechanism supports the intersubjective dimension of our subjectivity at the sub-personal level, providing a new cognitive dimension that assists us in defining our nature.

But what type of intersubjectivity do mirror neurons propose? They provide us with a new notion of intersubjectivity, the principal trait of which is intercorporeality. We will see how the motor system, together with its connections to the visceral-motor and sensory cortical areas, not only organizes the execution of the action, but also how it is perceived, imitated and imagined.

---

[4] Di Pellegrino et al. (1992); Gallese et al. (1996); Rizzolatti et al. (1996).
[5] See Gallese et al. (2004), and Rizzolatti and Sinigaglia (2006).

When the action is observed or imagined, its performance is inhibited. In this case the motor cortical circuits are activated, although not completely (certain components remain dormant) and not with the same intensity; hence the action is only simulated and not performed.

The discovery of the mirror mechanism both in animal (birds and non-human primates) and human brains opens a new evolutionary scenario that acknowledges "motor cognition" as a cardinal element for the appearance of human intersubjectivity.[6] We do not necessarily need to metarepresent the purposes and motor intentions of others in linguistic form in order to understand them. Most of the time we do not explicitly attribute intentions to others; simply, we just understand them. When we are present while others are doing something, we immediately comprehend most of their sensory-motor and emotional intentions without the need to explicitly represent them linguistically.

A whole new concept of human perception and cognition derives from this; intercorporeality becomes the primary, albeit not unique, source of our understanding of others. Our thesis is that these very mechanisms, situated within the framing and indexing that characterizes our relationship with the world of narrative fiction, also conflate our involvement with the stories we watch unfolding on the screen.

We will also see that action is just one of the vast range of experiences involved in interpersonal relations, including those regarding the characters in a movie. In fact, every interpersonal relationship implies the sharing of a multitude of states such as experiencing emotions and sensations, based on the parallel responses from inside our body, which are known as "enteroception."[7] Now we know that certain cerebral regions involved in subjective experiences such as touch or pain and emotions such as disgust or fear activate when we recognize them in other people. It follows that a vast number of mirror mechanisms

---

[6] Gallese (2000); Gallese et al. (2009).

[7] We are not able to give "affective neuroscience" the space here that it deserves. This field has helped us to understand that all our perceptions of the external world have a corresponding hedonic/affective experience that conditions our assessments, even those which, on the surface, appear objective and rational. We always observe the world from a subjective standpoint. Subjectivity cannot be attributed exclusively to the explicit formation of a judgment. In the first place, subjectivity means seeing the world from a perspective positioned in time and space, constituted by corporeal motor potentialities and always conditioned by the relationship with the world. In other words, it is characterized by an emotional experience, which can be pleasurable or not as the case may be, and accompanied by the body's variable internal conditions such as heart beat, blood pressure, hormones, etc. As a result of these internal and external bodily conditions, subject and object can be reduced to the verbal descriptions of the correlative aspects of a basic framework: The intentional relationship. For further reading on this subject, see the following authors (who are not always in agreement): Damasio (1994, 1999, 2010); Damasio and Carvalho (2013); Panksepp (1998, 2012); Solms and Panksepp (2012); Craig (2002); Tsakiris (2010).

are present in our brain. According to our hypothesis, these mechanisms allow us to recognize others as similar to ourselves and therefore make possible a first level of non-linguistic interpersonal communication and comprehension through the creation of an "intentional consonance."[8]

This, of course, does not mean that we are able to understand others *tout court* due to mirror mechanisms and simulation, or that we have a crystal ball in which to read their thoughts. It means there is no mystery to their actions and experiences as we share our corporeal nature and the underlying corporeal representational format at neural level. We have access to the "interior" of others; this does not pose a threat to their otherness, but does allow us a certain degree of comprehension. Intersubjectivity means that we are able to conjugate the dimension of otherness with that of identity. Our relationship with cinematographic make-believe is characterized by the activation of many of these mechanisms.

The functional architecture of embodied simulation appears to constitute a fundamental characteristic of the working of our brain, strongly implicated in our capacity for empathy. Empathy! This is a subject that alone could fill a whole book,[9] but we must limit ourselves to a brief overview. The notion of empathy was first mentioned in Germany in the second half of the nineteenth century during a debate on aesthetics. In 1873 the German philosopher Robert Vischer published his *Über das Optische Formgefühl: Ein Beitrag zur Aesthetik* ("On the Optical Sense of Form: A Contribution to Aesthetics"), in which he distinguished between the merely perceptive process of seeing and the pragmatically active process of looking.

Aesthetic enjoyment of objects/images in general and of works of art in particular implies an empathic involvement that manifests in the observer in a series of physical reactions. The observation of particular forms gives rise to reactive emotions, depending on their conformity to the design and function of the body muscles. The symbolic form, which is far from being pure as it is transcendental in a Kantian manner, derives its nature initially from its anthropomorphic content. According to Vischer, we manage to establish an aesthetic relationship with an object through a non-conscious projection of the image of our body. A few years later Theodore Lipps was to transfer this logic of the *Einfühlung* to the dominion of psychology and interpersonal relations; it was subsequently to have a significant influence also on Freud.

---

[8] Gallese (2006).
[9] We recommend Pinotti (2011) as an introduction for those interested in exploring in greater depth the concept of empathy, its history, and possible areas of application. See also Coplan and Goldie (2011).

Vischer's work also had a strong impact on two other persons of note: The sculptor Adolf von Hildebrand and the art historian Aby Warburg. In 1893 Hildebrand published a book entitled *Das Problem der Form in der Bildenden Kunst* ("The Problem of Form in Figurative Art"), in which he holds that the perception of the spatiality of the image is the result of a constructive sensory-motor process. In Hildebrand's view, space is a product and not a priori of experience as Immanuel Kant suggests. He affirms that the reality of the artistic image lies in its effectual nature, conceived as both the result of the causes that produced it and as the effect that it produces in the observer. According to this constructivist logic, the value of a work of art lies in the ability to establish a relationship between what the artist had in mind and the observer's reconstruction of the artist's intention, thus establishing a direct relationship between the creation of the object and the artistic pleasure it produces. According to Hildebrand, knowing the object is equivalent to knowing the process by which it was created. Another of Hildebrand's ideas is even more in line with our theory; he suggests that our experience of observed images has fundamental connotations in motor terms.

As Andrea Pinotti wrote in his valuable presentation to the Italian edition of Hildebrand's work:

> ...in Hildebrand's view everything begins with hand and eye movements; that is, when the body extends toward the construction of space. [...] It is movement which permits articulation of meaning, which allows connection of the elements available in space, it is movement that permits the formation of the object, representation and interpretation. [...] This is the reason why a work of art always contains indications of mobility, because it itself is a product of that mobility and at the same time requires the observer to activate his perceptive processes in order to deconstruct/reconstruct the image.[10]

Aby Warburg was a very perspicacious reader of Hildebrand's works. The definition of art historian is hardly adequate for this eclectic luminary,[11] whose works are still valid in today's context and worth studying in many aspects, also in the light of the hypothesis we are illustrating in this book.

A voracious reader, from Charles Darwin to his contemporaries, including the physiologists Hermann von Helmholtz, Ewald Hering, and Richard Semon, Warburg had no time for the disciplinary barriers which, lamentably,

---

[10] Pinotti (2011, pp. 15–16): Our translation.
[11] See Severi (2004) and Gallese (2012).

still hinder dialogue between life sciences and human sciences. He perceived the history of art as a means of clarifying the historic psychology of human expression and was convinced that the methodological frontiers of the study of art should be extended so that the history of art would be at the service of "a psychology of human expression that is still to be written."[12]

It is difficult to imagine how Warburg would have developed the concept of *Pathosformel* if he hadn't studied Darwin's *The Expression of the Emotions in Man and Animals* so meticulously during his sojourn in Florence. He didn't find a re-proposal of a rigid taxonomy of physiognomic expressions in Darwin's book, but he did find references to the role played by the central nervous system in directing the involuntary execution of corporeal gestures expressing a particular emotion. He also found references to the role played by habit in associating a given expression to a given emotional state, notwithstanding the apparent uselessness of such an association from a biological point of view. Finally, thanks to Darwin, Warburg discovered the biological need for a corporeal expression of emotions, transmitted in the form of non-conscious memory.

Warburg used the notion of *Prägung* (imprint) to characterize the survival of particular corporeal gestures and postures in the history of art. The delicate draperies, sinuous body movements, and curling locks of hair floating in the wind that characterize Sandro Botticelli's paintings are not only and exclusively the result of a conscious reproduction of the classical models; they are also and much more significantly proof of the survival of imprints of human expression (*Ausdrucksprägungen*). Indeed, Warburg, resolute in his intent of crossing the boundaries that separate the various disciplines, considered the history of art as a means of throwing light on the typically human power of expression. In so doing, he extended the methodological frontiers of the study of art in a completely innovative manner, opening the history of art to the contributions of science. This is just one of the reasons that Warburg's writings should be carefully reappraised today.

Another important contribution from this eclectic scholar is the concept of *Nachleben*, the survival of the individual and collective non-conscious memory of images and symbols, a concept with strong psychoanalytic implications. Here too Warburg was greatly influenced by the work of his two German contemporaries, the physiologist Ewald Hering (1834–1918) and Richard Semon (1859–1918), a biologist specialized in evolution. In Hering's work he found the notion of memory as the universal function of organized matter;

[12] Cited in Severi (2004).

memory includes the involuntary activation of images, sensations, and efforts. According to him, the central nervous system of every organism keeps track of its own past experiences that are then transferred to their offspring, a precursor of what is discussed today in epigenetics. The involuntary and non-conscious aspects of memory as described by Hering were crucial for Warburg's perspective on the history of art and the relationships between Antiquity, the Middle Ages, and the Renaissance. In fact, Warburg applied the same holistic and neurophysiological perspective to culture that Hering did to biology.

In Semon's work, however, Warburg discovered the notions of mneme and engram. For the biologist, the concept of mneme describes the ability of organisms to recall their experiences by means of the stimuli they encounter in their environment. This means that past and present are neurobiologically connected. He also introduced the concept of engram, the material trace recorded in memory by a variety of physical and psychological phenomena. Warburg used this concept to provide an energetic description of images, often referred to as "dynamograms"—energetic motifs. The serpents in the Laocoön are a dynamogram, a dynamic symbol that occupies the collective memory due to the energetic charge it has accumulated. With a significant overlapping of interests with the contemporaneous chronophotographic studies at the Collège de France in Paris conducted by Etienne-Jules Marey, one of the pioneers of cinema, Warburg saw the human figure as an energetic epiphany transformed into a body.

We believe that to the extent that cognitive neuroscience embraces the reflections and considerations of scholars of the caliber of Hildebrand and Warburg,[13] it can now provide a valid contribution to the study of human expressivity and aesthetic experience, refuting an approach that would reduce artistic creativity and the appreciation of aesthetics to the problem of where they are located in the brain. Through investigating the essential role of the body in creative expression and its reception, contemporary cognitive neuroscience has the concrete possibility of reviving this fertile line of thought that has been obscured for far too long by classic cognitivism's univocally abstract conception of nature and human intelligence, which, as time goes by, becomes less and less sustainable in the light of the new discoveries emerging as neuroscience interrogates the brain–body system.

---

[13] Space does not permit an analysis of the equally important contribution of philosophic anthropology to the study of human creative expressivity. We would draw the reader's attention to the writings of Viktor von Weizsäcker and his concept of *Gestaltkreis*, which holds that the reciprocity between movement and perception intuits the temporality and spatiality of the event. The sensory–motor relationship is therefore interpreted in morphogenetic terms. See Tedesco (2008).

When we talk of embodied simulation, we are asserting that body parts, actions, or more in general corporeal representations play a critical role in the cognitive processes. Mental states or processes are embodied insofar as they are represented in corporeal form. An action or a motor intention can be represented in either corporeal or linguistic form.[14] We do not know if and to what extent the linguistic representation is totally separated/separable from corporeal representation; however, empirical data seem to point to it not being so.[15]

The concept of embodiment might be erroneously construed as mind pre-existing the body and then using it as a form of domicile. In point of fact mind and body are two different levels of description of the same reality, externalizing different properties depending both on which level of description has been selected and the language used to describe it. Thoughts and ideas, perceptions, and mental images are, of course, neither muscles nor neurons, but their contents are inconceivable if severed from our situated corporeality. We also employ forms of representation that use non-corporeal formats; it is widely held that we do this when we use language, for example. It is, however, difficult to imagine how a representational linguistic format could have developed separately from our corporeality. It appears that we can transcend it with language, but numerous empirical data suggest that there is always a link with the body.

Embodied simulation coincides with the concept of empathy in many respects, but, in fact, goes further. As we will see later in this chapter, embodied simulation underlies important aspects of the construction of our spatial maps, influences our relations with objects, and forms the basis of our faculty of imagination. Later in this book we will show how these aspects are relevant in explaining involvement in cinematographic make-believe and the fact that they cannot simply be categorized as expressions of empathetic competencies. We fully agree with Gregory Currie that the notion of (embodied) simulation embraces a wider range of aspects than empathy in throwing light on our involvement in works of fiction, particularly films.[16]

In our view, embodied simulation could constitute a fundamental point of access to the world, providing genetic support to the various levels through which we conceive reality, unifying at corporeal level our experiences of the

---

[14] Alvin Goldman made significant contributions to the concept of the representational corporeal format. See Goldman and de Vignemont (2009), and Goldman (2012).
[15] See Gallese and Lakoff (2005); Gallese (2008), Pulvermüller and Fadiga (2010); and Glenberg and Gallese (2012).
[16] Currie (2011).

myriad worlds that impinge on our own, including those of fiction, both artistic and not.[17]

Before discussing the empirical data that support our theory, we will dedicate a short excursus to the brain–body relationship and the specificity of the neuroscientific approach we are proposing to apply to it.

Body, brain, and neuroscience

The second half of the twentieth century saw significant progress in neuroscience, due also to the recently developed brain imaging techniques, such as functional magnetic resonance (fMRI), that has made it possible to study the brain from new standpoints without incommoding its owner.

Neuroscience started investigating fields such as intersubjectivity, self, empathy, decision-making, ethics, aesthetics, economy, and much more besides. This gave rise (and still does) to doubts as to the legitimacy and/or capacity of the discipline to throw new light on certain characteristic aspects of human subjectivity such as art, creativity, aesthetics, politics, and, last but not least, the distinctive traits of human nature.

But what exactly is cognitive neuroscience? First and foremost, it is a methodological approach, whose results are strongly influenced by the assumptions of the theoretic framework of reference. The study of an individual neuron and/or the brain itself cannot anticipate the questions, even less the answers, that this scientific approach to understanding the workings of the human mind can posit.

This chapter provides a brief introduction to the key investigative methods currently in use by neuroscience, highlighting their strengths and weaknesses. We will then propose an alternative approach, illustrating its advantages from both the methodological standpoint and that of the results that it can guarantee. Our starting point for this approach is the assumption that the study of the brain alone, if detached from the analysis of the close relationship of interdependence between body, brain, and the surrounding environment, does not suffice for a study of human nature, particularly not if the objective is to throw new light on the issue of intersubjectivity and its mediated forms such as the aesthetic experience of images in movement, one of the basic aspects of watching a movie.

---

[17] Freedberg and Gallese (2007); Wojciehowski and Gallese (2011); Gallese and Guerra (2012, 2013a, 2013b).

Our body represents our main source of pre-reflective self-awareness and awareness of others and is also the root and the base on which all forms of explicit and linguistically mediated cognition of objects are developed. Seen from this standpoint the body is a priori the ultimate source of our experience of ourselves and our relationship with the world. When we talk about the body in this context, we are attributing it a double role and two complementary and tightly entwined modalities, *Leib* and *Körper: Leib*, our self-experience, experience of others, and the world that surrounds us; *Körper*, the material body, studied by physiology, and of which the brain, studied by neuroscience, is an integral part.[18] For this reason, when we refer to the brain during this excursus we are always referring to the *brain–body system*.

We suggest that the dual nature of the body and its genesis can best be examined and understood starting from a neurophysiological study of its subpersonal sensory–motor and affective parts. The naturalization of intersubjectivity, including its mediated forms such as the experience of film, should start from the deconstruction and reconfiguration of the concepts that we normally use to describe it by literally studying their component parts and where they come from through an investigation and the level of description of the brain–body system as in neuroscience. The deconstruction of these concepts is therefore an investigation of their constitutive mechanisms. We will illustrate the results of neuroscientific research that attribute a new meaning to the concepts of purpose, space, object, action, emotion, and sensation. This review of the neurophysiological mechanisms that underpin important aspects and dimensions of these concepts provides the foundation for the formulation of a new model of perception of images, including those we see on film, and how it produces experiences in the observer.

Our proposal includes another important aspect: The comparative perspective. This provides the opportunity of inserting the theme of perception of images in an evolutionary framework that studies its phylogenetic antecedents. Of course, this does not mean we are refuting the indisputable difference of human intelligence compared to that of other animal species; simply we consider that a comparative perspective greatly reduces the risk of the study of the human brain being subordinated to a predetermined and specific model that has been taken as being valid a priori. Another reason for privileging this perspective is that it offers the possibility of using more efficient and specific empirical research tools characterized by a sharper spatial–temporal resolution,

---

[18] We have adopted the distinction proposed originally by Edmund Husserl (1931, p. 119).

which can correlate the activity of the individual neurons with various aspects of behavior and cognition.

From classic cognitivism to embodied cognition

Vast areas of neuroscience are still strongly influenced by classic cognitivism on the one hand and by a number of theoretical aspects proposed by evolutionary psychology on the other. Classic cognitive science is characterized by a solipsistic vision of the mind, in which the social dimension of human existence is merely the arena in which the rationality of the abstract and disembodied mind of the individual unfolds. Put in more explicit terms, this means that our actions, our creative expressions, our sensations and emotions have meaning for us only to the extent that they are the object of a linguistically precise representation. According to this model, the reception and understanding of actions, creative expressions, sensations, and emotions of others would comply with the same rules. The theory of the mind that classic cognitivism offers is that of a functional system that manipulates abstract amodal symbols complying with formal syntactical rules. Taking these premises as a starting point, neuroscience was (and partly still is) induced to study the brain, searching for a hypothetical one-to-one map between the concepts used to describe the human mind, its functions, their supposed cause, and neural origins.

A second salient aspect of classical cognitivism that has influenced neuroscientific research for many years is the relationship between action, perception, and cognition. According to the classical cognitivism model that Susan Hurley criticized using the "classical sandwich" metaphor (Fig. 1.2),[19] action and perception are two distinct faculties, produced by two specific and equally distinct cerebral circuits. Moreover, according to the model they are peripheral to the cognitive processes which Hurley intentionally describes as the sandwich filling.

There is a third aspect; the unidirectional information flow connecting perception, cognition, and action. The sensory apparatus supplies input for a perceptual representation of the world, characterized by the association of the individual sensory modalities run by the so-called associative areas, which are considered hierarchically superior to the visual, somatosensory, and acoustic unimodal areas. The associative areas supply external input to the cognitive apparatus, which, after having processed the information, proceeds to influence

---

[19] Hurley (2002).

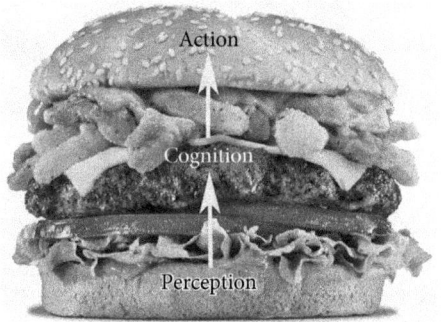

Fig. 1.2 The classical cognitivism model of the relationship between perception, action, and cognition, depicted as a sandwich.

the world by activating the motor system and triggering movement and actions. To all extents and purposes, the model relegates the motor system to the lowly role of movement controller, devoid of any perceptual function and peripheral to the cognitive processes. In other words, what we do and what we see in real life and on the screen, be it cinema, television, or any other form of digital support, are simply the individual products of an abstract computational cognitive system. The old dream of translating the world into numbers or even conceiving it as the material expression of an ideal pre-existent symbolic/numerical model, reconfigures it in a neuroscientific approach to the human mind that not only loses sight of the brain–body link, but totally ignores the experiential dimension of human existence. In our view this exclusive and morbid clinging to language to explain our relationship with the external world and the worlds of fiction excludes the possibility of a clearer understanding, as it explains the relationship with the world exclusively or prevalently in terms of formulations of hypotheses and the application of deductions and logical inferences. Moreover, the classical cognitive model is also refuted by our daily experience, although we are well aware that this cannot be considered an argument from the scientific point of view. Later in the book we will see how this model has been challenged in recent decades by empirical neuroscientific research.

Evolutionistic psychology holds that the human mind consists of a collection of cognitive modules, each of which has been selected over time for its adaptive value. In this case the relevant metaphor is the Swiss army knife (Fig. 1.3). This extremely useful instrument is composed of a number of utensils, including blades, scissors, screwdrivers, bottle openers, files, and so on, each

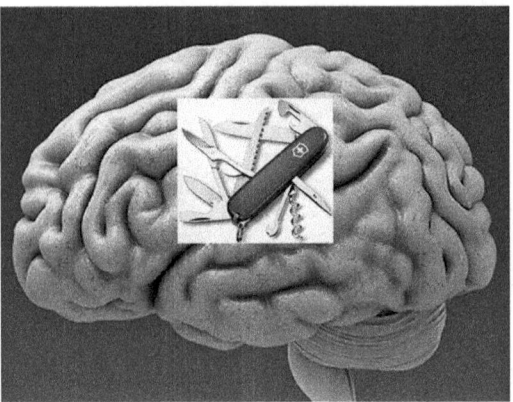

Fig. 1.3 Metaphor of the Swiss Army knife illustrating the supposedly modular organization of the brain.

of which has a very specific function; according to evolutionary psychologists, this reflects the structure of the mind, which, in the course of its evolution, has acquired a series of cognitive modules, each delegated with a particular cognitive or mental function.

Leading supporters of this line of thought, such as Leda Cosmides and John Tooby, maintain that the brain is a physical system that functions like a computer.[20] Steven Pinker says that our cognitive life sources form the function of a series of modules, such as the linguistic module, the module for the Theory of Mind, and so on.[21]

Taking this theoretical framework as a basis, in recent times many neuroscientists have concentrated on identifying the site of these "cognitive modules" in the human brain.[22] What emerged is an approach characterized by a sort of neo-phrenological ontological reductionism, which transforms the individual and his mental faculties into a collection of cerebral circuits and areas, each of which is hypothetically characterized by a particular psychological or cognitive function and a specific adaptive value.

However, this ontological reductionism is also vitiated by an excessive faith in brain imaging visualization techniques (such as fMRI), frequently considered to be the only research method available. fMRI methodology studies

---

[20] Cosmides and Tooby (1997).
[21] Pinker (1994, 1997).
[22] See Gallese (2014); Ammaniti and Gallese (2014); and Gallese and Cuccio (2015) for a recent criticism of classic cognitivism and evolutionary psychology.

the functioning of the brain only indirectly, correlating the increase in local blood flow and the hypothesized increase in neural activity; it raises the activation of individual cerebral areas or, at best, the activation of circuits that connect different areas and regions of the brain, to a descriptive level. Moreover, the technique is hampered by the low signal-to-noise ratio, due to the distortion and intrinsic noise inherent to indirect measuring of cerebral activity, a problem generally dealt with by taking measurements from more than one person and then calculating the average level of activation during the research task. Hence, if fMRI is to be used in an experiment, 15–20 participants will be needed in order to calculate the average cerebral activity. With the tool's spatial resolution, it is possible to estimate indirectly the activity of the neurons contained in a voxel measuring just a few millimeters per side for each brain. In the best scenario, the activity of no fewer than several hundreds of thousands of neurons are estimated, without the possibility of determining whether their activity is excitatory or inhibitory. The temporal resolution is even less effective; from the moment it starts to measure variations in the local blood flow to the brain, it is only able to identify cerebral activity for a period of no less than a few seconds. Unfortunately, the individual action potentials, the electric charges emitted by active neurons and used by them to "intercommunicate," last less than a millisecond. We certainly do not intend to minimize the important results obtained by fMRI; however, we would point out that when the data obtained from these studies, such as those that clarified the organization of the visual cortex, face recognition, memory mechanisms, and so on, were inspired by and/or interpreted in the light of data previously obtained through the recording of individual neurons in non-human primates and other animals, they turned out to be more pertinent and reliable.

Fortunately, fMRI is not the only experimental methodology available for studying the human brain. There is also high-density electroencephalogram (EEG), the tool we ourselves used in the experiments with films that we will illustrate later; magnetoencephalography (MEG) with its temporal resolution of milliseconds and therefore very similar to that of the recording of individual neurons; or transcranial magnetic stimulation (TMS) for studying the effects of the stimulation or temporary and reversible inactivation of very small cortical regions (just a few millimeters in diameter). When used in repetitive mode to inactivate the cortex, TMS can establish a causal relationship between the functionality of a given cerebral region and a certain perceptive or cognitive executive function. From this it is possible to analyze the fMRI results, which are only correlative, and formulate causal hypotheses regarding the functional role of the cerebral circuits and areas.

Another possible area of research is offered by patients who have suffered cerebral damage or lesions. Scientific studies of their symptoms have provided, and continue to provide, a basis on which to formulate hypotheses regarding causal relations between lesion sites and the functions altered or destroyed by the lesions themselves. However, the research technique that has most facilitated progress in the understanding of cerebral functions is the recording of the individual neurons. This technique, in which one of the authors has specialized, is used prevalently with animals such as mice, rats, and monkeys, correlating their neural activity with various aspects of their behavior. In a limited number of cases it can also be applied to the study of human neurons, though exclusively for clinical reasons, and the results obtained are also useful for neurocognitive research.

The use of this technique in studies of the brain of the macaque monkey has greatly increased our knowledge of the organization of the sensory and motor systems, memory, reward, and decision mechanisms. What is truly astounding is that many of the discoveries made regarding the brains of non-human primates by recording individual neurons have subsequently been found in the human brain, even using methods that "see" directly or indirectly the activity of an enormously superior number of neurons. The discovery of canonical neurons and mirror neurons is an example of this.[23]

The approach we are proposing here advocates the necessity of accompanying brain imaging with a detailed phenomenological analysis of the perceptive, motor, and cognitive processes to be studied and—even more importantly—of interpreting the results in the light of the deficits caused by cerebral lesions studied by clinical neuropsychology, and on the basis of the activity of the individual neurons in animals and in the non-human primates.[24] In other words, if neuroscience is simply declined as brain imaging, it will lose a great deal of its heuristic power, which is even further diminished when accompanied by the instrumental use of empirical data to corroborate a preconceived model of disembodied mind—the classical cognitivism model—which is taken as valid a priori.

All the same, we are witnessing a radical change in the paradigm. A new neuroscientific approach to the study of the human condition, known as embodied cognition,[25] is indeed possible and is progressively gaining consensus.

---

[23] See Gallese and Cuccio (2015) for further reading on this subject.
[24] Gallese (2001, 2003).
[25] There are, in fact, various ways of conceiving cognition as being "embodied," but space does not permit an in-depth discussion of them in this book. Recommended further reading: Lakoff and Johnson (1980, 1999); Lakoff (1987); Goldman and De Vignemont (2009); Goldman (2012); Uithol and Gallese (2015).

Our hypothesis is perfectly aligned with this approach and, moreover, stresses the need to focus research on our relationships from the point of view of experience, applying a form of "phenomenologization" of cognitive neuroscience.[26]

An analysis of the role that our body plays in building experiences of objects and other people, both in real life and in the realms of narrative make-believe, could provide a new basis for an empirical study on the constitution of subjectivity and intersubjectivity compared to that used by classic cognitivism, i.e., the first-person aspects of experience would be retained. Francisco Varela had already ventilated this possibility as far back as 1999, and, indeed, started a research study in this direction.[27]

The objective is to demonstrate how, starting from a sub-personal neuroscientific description of our pragmatic and affective relationship with the world around us, it is possible to plot a research program based on the biological and anthropological dimension of the human being that defines the distinctive forms of subjectivity and intersubjectivity and the resulting expressive and receptive modalities of these creative and symbolic expressions.

The next section illustrates the recent contributions of neuroscience to achieving a radical modification of our action, perception, and cognition models, a change that has challenged the classical sandwich model from which we started.

Motor cognition: Movements and motor goals

For many years it was thought that the frontal lobe was divided into three parts: The posterior section containing the primary motor area, previously identified also as Brodmann's area 4, but now known as F1; then came a premotor area, Brodmann's area 6; and in front of that the area known as the prefrontal lobe. According to this model, only the processing of the information in the prefrontal portion is cognitive in nature.

Things are very different today (Fig. 1.4): Area 6 as such no longer exists. In fact, it includes a mosaic of regions which are cytoarchitectonically different and are also distinct from the point of view of connectivity with other cerebral areas. Each of these regions of area 6 has been allocated a designation

---

[26] On this topic, see also Gallese (2001, 2007, 2011), and Gallese and Cuccio (2015). Murray Smith stressed that our understanding of the conscious experience of the real world (and the world of cinema) will be able to progress only if there is a triangulation between the phenomenological, psychological, and neural levels of description (Smith, M., 2012, p. 88). See also Flanagan (1992).
[27] Varela and Shear (1999).

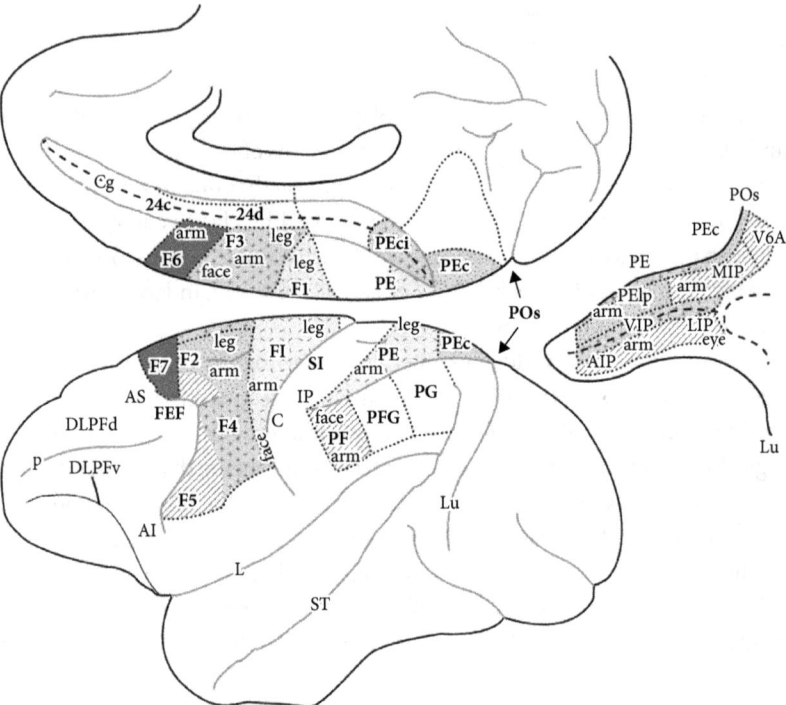

Fig. 1.4 Lateral and mesial views of the brain of a macaque, showing the parcellation of the motor cortex, the posterior parietal cortex, and the cortical cingulate areas, with the principal somatotopic arrangement. The drawing on the right-hand side of the diagram shows the cortical areas buried in the intraparietal sulcus. The reciprocally connected motor and parietal areas are indicated by the same graphic symbols. The motor areas that mainly receive input from the prefrontal areas are shaded in dark grey.

AI, inferior arcuate sulcus; AIP, anterior intraparietal area; AS, superior arcuate sulcus; C, central sulcus; Cg, cingulate sulcus; DLPFd, dorsal dorsolateral prefrontal cortex; DLPFv, ventral dorsolateral prefrontal cortex; L, lateral fissure or fissure of Silvius; LIP, lateral intraparietal area; Lu, lunate sulcus; MIP, medial intraparietal area; P, principal sulcus; POs, parieto-occipital sulcus; ST, superior temporal sulcus.

Reprinted from *Neuron*, 6 (13), Giacomo Rizzolatti and Giuseppe Luppino, The Cortical Motor System, pp. 889–901, Figure 1, doi.org/10.1016/S0896-6273(01)00423-8 Copyright © 2001 Cell Press. All rights reserved. With permission from Elsevier.

beginning with F and followed by a progressive number; each entertains reciprocal connections with specific regions of the posterior parietal lobe, the cerebellum, the basal ganglia, and other areas of the prefrontal portion of the frontal lobe. What is even more important is that each of these regions contains neurons with different functional properties.

What does the motor system actually do? For many years the answer was as simple as the question itself, it produces movement. Today we know that this is not the correct answer, or at least it is only partially correct. The motor system certainly does produce movement, but it also produces motor acts and actions; in other words, movements with a goal, a purpose such as grasping an object, or a series of movements to achieve a more remote goal, such as grasping a mug and lifting it to the mouth to drink from it. To summarize, a movement is a simple dislocation of body parts, like flexing and stretching fingers. A motor act, however, consists of using these movements to attain a motor goal, as when grasping, holding, manipulating, breaking, or placing an object in a given position.

A fundamental contribution to this area of research has been made by Giacomo Rizzolatti and his team of the University of Parma, starting from the early 1980s. While experimenting with F5, the premotor ventral area of the macaque, they discovered that the neurons in the area did not activate during simple movements, but only when a motor act was involved. The same motor neuron activates independently of whether the macaque grips the object with its right paw, left paw, or even with a completely different effector such as the mouth.[28] The common denominator that activates these neurons is neither muscle nor movement, it is the goal behind the motor acts. Premotor neurons, which were traditionally thought to be part of the common final step by which the agent executes movements in response to external or self-generated stimuli, are now revealed to be correlated to the highest abstract level of description of movement: Its finalism.

The teleological nature of the motor neurons, in other words the fact that they relate to actions with a finality—the achievement of a goal—rather than to movement as such, has been definitively demonstrated with an experiment illustrated in detail in Figure 1.5.

This experiment shows that the F5 motor neurons fire independently of whether the monkey is using normal tongs to grip the object, when the tongs are closed by closing the fingers around the arms of the tool, or inverted tongs, when the object is gripped by opening the fingers. The movements are different, but the goal of gripping the object is the same. The premotor neurons are much more sensitive to the goal behind the action than to the individual movements necessary to attain it. This is an extremely important finding as it opens the way to a radical rethinking of the motor system in cognitive terms.[29]

---

[28] See Gallese (2000) and Rizzolatti and Sinigaglia (2006).
[29] Umiltà et al. (2008).

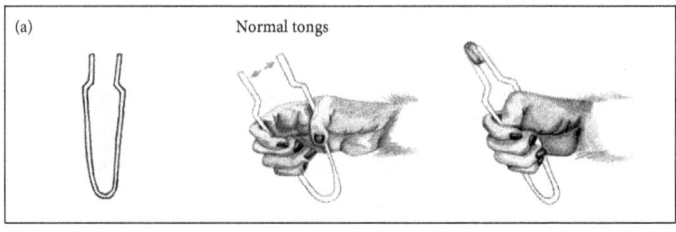

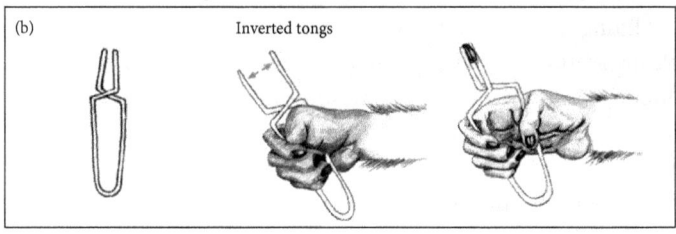

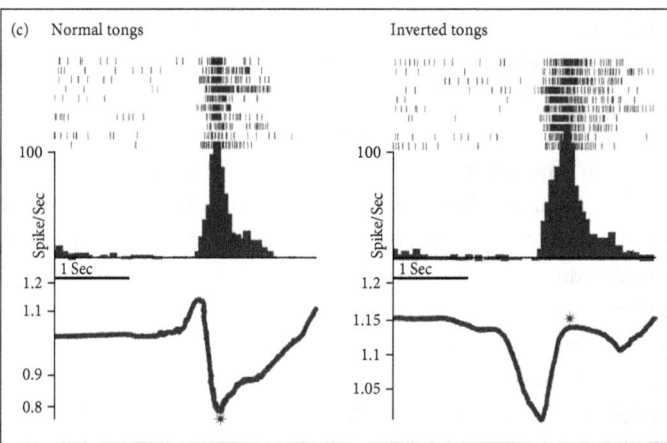

Fig. 1.5 Schematic representation of the experiment. In order to grip the object with (A) the normal tongs, the monkey has to close its fingers, while to grip the object with (B) the inverted tongs, it has to open them. (C) The activity of a motor neuron registered by area F5 while grasping an object with normal tongs (left) and inverted tongs (right). The individual action potentials (spikes) and the histograms below are aligned with the final phase of the gripping with the tongs (indicated by an asterisk). Time is recorded on the abscissa axis, while the ordinate axis shows the intensity with which the neuron fires, measured in spikes per second. The lines below the histograms show the position of the fingers, measured in volts by potentiometers as distance between the prongs of the tongs: The descending section of the line indicates the closure of the fingers and the ascending section, the opening of the fingers. In both conditions the neuron reaches the maximum response when the grip is completed, independently of whether the fingers are open or closed.

Reprinted from Umiltà, M.A., L. Escola, I. Intskirveli, F. Grammont, M. Rochat, F. Caruana, A. Jezzini, A., V. Gallese, G. Rizzolatti (2008), "How pliers become fingers in the monkey motor system", *PNAS*, 105, pp. 2209–2213.

Motor cognition: Area F4 and peri-personal space

The first form of embodied simulation that we will examine closely regards the way in which the brain–body system maps the space around the body, known as peri-personal (or near) space.[30] This hypothesis is based on a series of results obtained by recording the properties of the premotor F4 neurons. F4 is located immediately in front of the primary motor area (F1) and behind F5, which was discussed in the previous section. Many F4 neurons control the motor acts of the arms, such as reaching for an object in peri-personal space, and of the trunk and head, such as moving towards an object in peri-personal space with the objective of reaching it or moving away from it in order to avoid it. It is interesting to note that as well as having these motor properties, the great majority of F4 neurons also react to other sensory stimuli. These neurons that control movement also fire in response to tactile stimuli applied to body parts such as the arms and head, visual stimuli, and acoustic stimuli from areas close by (Fig. 1.6).

The visual properties are shown as three-dimensional receptive fields limited to peri-personal space in correspondence with the tactile receptive fields. The neuron that controls the act of moving the arm closer to an object also responds to tactile stimuli applied to the arm, to visual stimuli in the area in which the arm is positioned or acoustic stimuli from close by. If the arm is moved, then the visual receptive field moves too, independently of the direction of the gaze (Fig. 1.7).

These visual fields are mapped with a system of somatocentric coordinates; in other words, they are centered on the corporeal parts, independently of the position of the eyes, remaining anchored to the underlying tactile receptive field.

The study of the properties of these neurons has revealed a high degree of congruence between the spatial position towards which the arm is directed when moving to reach an object and the localization of the visual receptive field.

The presence of a motor response and tactile, visual, and acoustic responses in a single neuron suggests that the contact, the sight or the sound of a stimulus close to the body, evokes the simulation of the reaching movement, oriented towards or away from the stimulus itself. What we want to convey here is the priority of motor space over visual space. Our hypothesis is that the vocabulary of reaching and orienting movements in space forms as a consequence of the

---

[30] Peri-personal and extrapersonal space are defined, respectively, as portions of space within hand's reach and out of hand's reach. See Rizzolatti et al. (1997).

22 EMBODIED SIMULATION

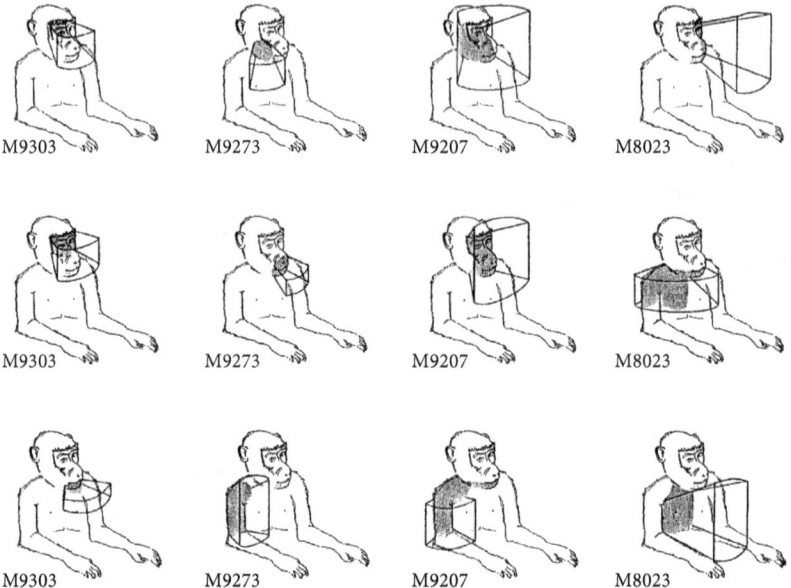

Fig. 1.6 Somatosensory and visual receptive fields of F4 bimodal neurons. The shaded areas indicate the tactile receptive fields; the shapes around the animal delineate the visual receptive fields.

Adapted from *Journal of Neurophysiology*, 76 (1), L. Fogassi, V. Gallese, L. Fadiga, G. Luppino, M. Matelli, and G. Rizzolatti, Coding of peripersonal space in inferior premotor cortex (area F4), pp. 141–57, doi.org/10.1152/jn.1996.76.1.141 Copyright © 1996 the American Physiological Society.

movements of the arm or head to reach objects in the surrounding space. This generates a motor space based on the strengthening of the connections needed for the action to be successful. Tactile, visual, and acoustic information settles into and takes form in this motor space. The simulation of the motor potentialities of the corporeal parts creates a motor space that represents the base, the a priori motor construction of the space surrounding the body that organizes, integrates, and gives meaning to the sensory information regarding the body and all that happens in its immediate vicinity.

In this case, too, successive research studies brought to light analogous properties in the human motor brain. It has been demonstrated that the ventral part of the human premotor cortex, similarly to the F4 neurons in the macaque, responds to tactile stimuli applied to the face and to visual and acoustic stimuli in the peri-personal space surrounding it.[31] In addition, temporary inactivation

---

[31] Bremmer et al. (2001).

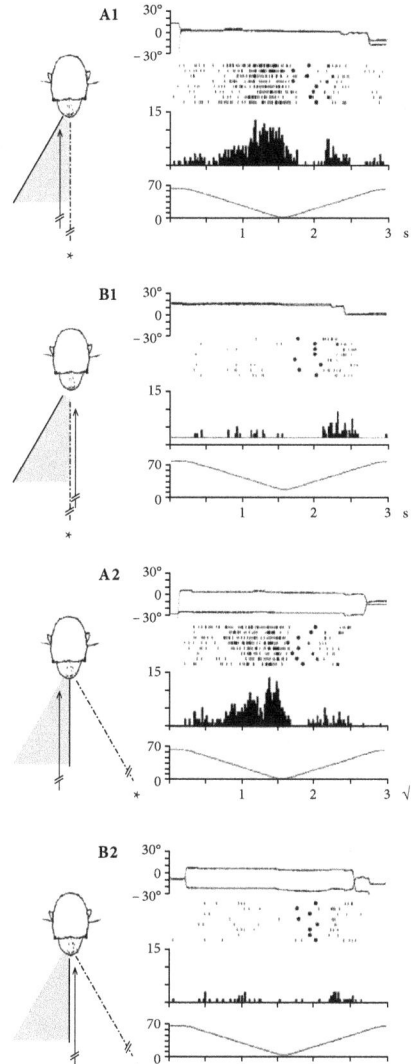

Fig. 1.7 Bimodal F4 neuron with the visual receptive field anchored to the body. From top to bottom: Horizontal and vertical eye movements, neural discharge during individual trials, the response histogram (abscissa axis: Time; ordinate axis: Spikes/bin, bin dimension 20 ms), and the variation over time in the distance between the stimulus and the monkey's head. The descending section of the curve indicates the movement of the stimulus towards the monkey, while the ascending section represents the movement of the stimulus in the opposite direction (abscissa axis: Time in seconds; ordinate axis: Distance in centimeters). The tactile receptive field is localized in the right hemiface. The visual receptive field is positioned around the tactile field.

Adapted from *Journal of Neurophysiology*, 76 (1), L. Fogassi, V. Gallese, L. Fadiga, G. Luppino, M. Matelli, and G. Rizzolatti, Coding of peripersonal space in inferior premotor cortex (area F4), pp. 141–57, doi.org/10.1152/jn.1996.76.1.141 Copyright © 1996 the American Physiological Society.

produced by repeated TMS of the premotor cortex interferes with the processing of multisensorial stimuli in the peri-personal space surrounding the hand. These results show that the cortical motor system maps the body's motor potential, which permits the integration of stimuli regarding the body and its surrounding space.

As anticipated by Maurice Merleau-Ponty in his *Phenomenology of Perception*, "my body appears to me as an attitude, directed towards an existing or potential task. And indeed, its spatiality is not, like that of external objects or like that of 'spatial sensations', a *spatiality of position*, but a *spatiality of situation*."[32] Peri-personal space is multisensorial (based on the integration of visual, tactile, acoustic, and proprioceptive information), centered on the body (controlled by a system of coordinates that are anchored to the various body parts and not to the position of the eye), and motor in nature. Once again Merleau-Ponty's words come to mind; peri-personal space and its extension can be constituted as "the varying range of our aims and our gestures."[33] In our view, the spatiality of the situations described in a movie, notwithstanding the space that divides the screen from the audience, can be experienced and perceptively understood also through the simulation of the motor potentialities of the actors and the movements of the camera recording them. This hypothesis will be discussed in depth in the following chapters of the book.

Motor cognition: Canonical neurons and objects "close to hand"

Now we will see how the perception of manipulable objects, similar to the perception of peri-personal space, is another example of embodied simulation in the field of action, which cannot be attributed to the notion of empathy in the strict sense. Many studies involving the recording of the individual neurons of the macaque have shown that the neuron population in the F5 premotor area and the posterior parietal anterior intraparietal area, which are reciprocally connected, activate selectively both during the act of grasping an object and when the monkey is merely observing the object. These neurons are known as "canonical" neurons.[34]

When we look at an object that can be manipulated in some way, the same motor circuits that are normally activated during the planning process and

---

[32] Merleau-Ponty (1945, pp. 114–15).
[33] Merleau-Ponty (1945, p. 166).
[34] Jeannerod et al. (1995); Murata et al. (1997, 2000); Raos et al. (2006); Umiltà et al. (2007).

the implementation of actions involving the object are recruited selectively (Fig. 1.8).

The most interesting aspect of the canonical neurons is that a good number of them shows a high level of congruence between the motor selectivity for a particular type of grip (e.g., positioning the index finger opposite the thumb to pick up small objects) and the visual selectivity shown for small objects which, albeit differing in shape (e.g., cubic, conical, or spherical), still need the same type of grip controlled by the same neurons. These visual responses do not prepare for or prelude any actions regarding the objects at which the macaque is looking. Similar brain imaging studies conducted later with humans returned analogous results. When a person sees an object that can be manipulated to achieve a result, such as a key, a glass, or a gun, the visual stimulus evokes a motor activation in that person's brain even if he has no intention of carrying out a grasping or gripping action. A recent study performed with humans has shown that the reversible inactivation of a limited portion of the ventral premotor cortex with repetitive TMS alters the perception and recognition of the shapes of objects.[35] These results reveal a very simple fact, which, however, has revolutionary implications: Our motor system activates also when we are not moving. When operating in this simulation modality, the motor system relates what we observe back to us, transforming/processing what we have seen into an interior understanding.

The visual response of the canonical neurons maps or represents objects in relational terms. When we look at an object and no active interaction is required, just the mere act of looking triggers the activation of the motor program that would have been used had we wished to interact with it. In fact, seeing the object means that we automatically simulate what we do with it; we are simulating a potential action. In other words, objects are not just identified, differentiated, and categorized on the basis of their physical appearance, but also by the effects of their interaction with a potential *agent*.[36]

In this perspective, the dynamic relation with the subject/agent endows the object with full significance and a meaning; the relation is undoubtedly multiple, just as the ways in which we interact with the world around us are multiple as we move, exploring objects with our eyes and other senses. Through embodied simulation, perceiving an object can be seen as a preliminary form of action which, independently of any physical interaction, presents it as being

---

[35] Uithol et al. (2015).
[36] We believe that this interpretation is not only congruent with the notion of affordance proposed by Gibson (1979), but also that it provides a more precise definition in neurophysiological terms as it

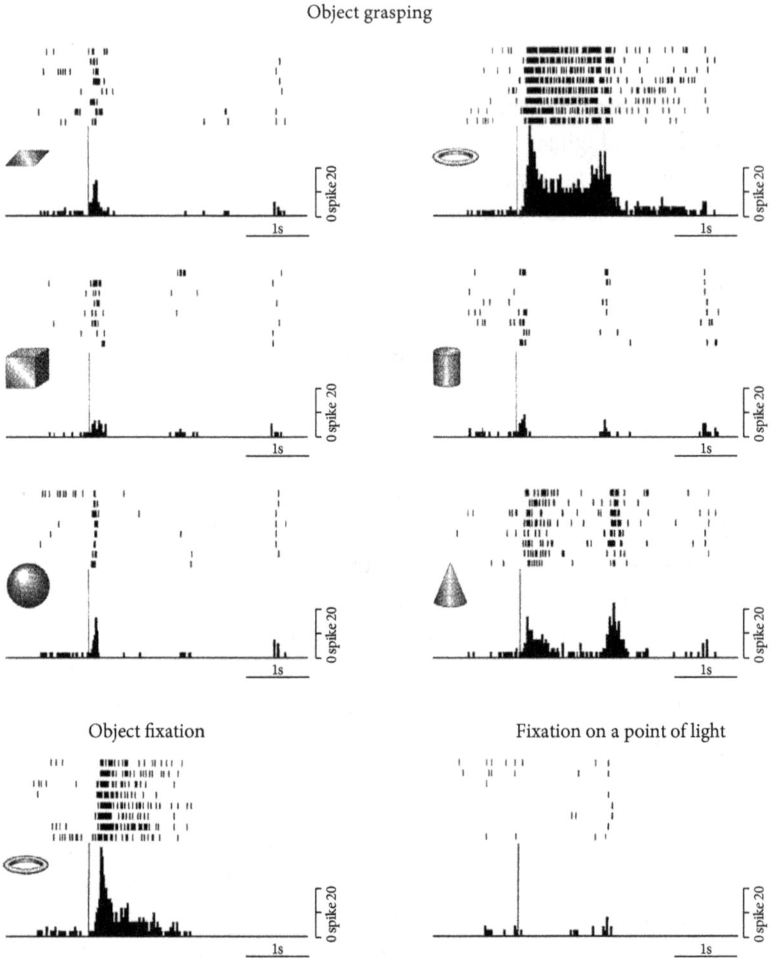

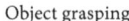

Fig. 1.8 The F5 canonical neuron. The upper section of the diagram illustrates the activity of the neuron while various objects are being observed and grasped in an environment in which the light is switched on and off. The rasters and the histograms are aligned at the moment in time when the monkey presses a lever and the object becomes visible. It is evident that the neuron responds selectively to the ring; there are two response peaks, the first being triggered by the sight of the ring and the second by the gripping movement. The visual response to the ring is also present in the experimental condition in which the monkey stares at it without having to pick it up (the bottom panel on the left). The bottom panel on the right shows the response of the neuron when the monkey cannot see any of the objects but is instructed to stare at a point of light.

Reprinted from *Journal of Neurophysiology*, 78 (4), Akira Murata, Luciano Fadiga, Leonardo Fogassi, Vittorio Gallese, Vassilis Raos, and Giacomo Rizzolatti, Object Representation in the Ventral Premotor Cortex (Area F5) of the Monkey, pp. 2226–2230, Figure 1, doi.org/10.1152/jn.1997.78.4.2226 Copyright © 1997, The American Physiological Society.

something "to hand" (*zu Handen*, in Martin Heidegger's words)[37] and gives it presence.[38] This suggests that embodied simulation shapes the content of the visual perception, characterizing the object perceived in terms of the motor acts it evokes even if there is no actual movement involved. The implications these results have on our experience when watching films will be examined in greater depth in Chapter 3.

Motor cognition: Mirror neurons and mirroring mechanisms

We have already spoken of area F5 regarding how the finalities of motor acts are included in the functional attributes of the motor system and how canonical neuron mechanisms map the simulation of motor acts at the sight of objects with which we interact. This premotor area also houses a population of neurons, initially discovered in the macaque, whose motor aspects are indistinguishable from the other neurons present in the area but whose visual properties are very different; these neurons are known as "mirror neurons."[39]

The principal characteristic of mirror neurons is that they activate both when a motor action is executed, such as gripping an object or forming communicative gestures with the mouth, and when observing someone else performing the same action or making the same gesture (Fig. 1.9).

In our view, mirror neurons are important not only because they were discovered in area F5 with the properties described earlier, but also and above all because they have led to the discovery of a functional neurophysiological mechanism that for the first time shows the connection between two individuals, mapping the actions of one onto the motor system of the other.

This discovery has supplied the first neurobiological correlate of the relevant aspects of *certain* philosophical and psychological theories, as well as various intuitions on the nature and functioning of the social relations between individuals. However, at the same time it has revealed the limits or implausibility of *other* philosophical and psychological theories, and intuitions on the same subject.

---

attributes a greater importance to the active role played by the motor potentialities of the body of the observer in the constitution of the pragmatic invariants specified by the object.

[37] Heidegger (1927).
[38] The relation between the brain and the concept of presence will be explored further in Chapter 6.
[39] Di Pellegrino et al. (1992); Gallese et al. (1996); Rizzolatti et al. (1996).

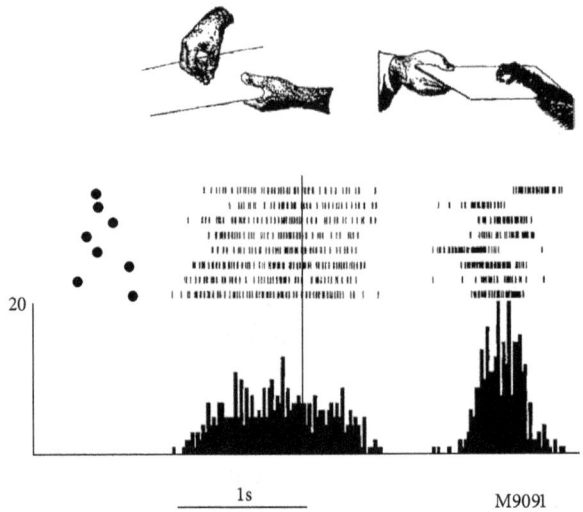

Fig. 1.9 Activation of a mirror neuron while gripping an object and observing the action of gripping an object as registered by area F5 of a macaque. The top panel shows the two conditions that activate the neuron: On the left-hand side, the gripping action is performed by the experimenter in front of the monkey, who simply observes the action; the diagram on the right-hand side shows the action performed by the monkey itself.
The bottom panel shows the individual spikes and histogram of the response of the neuron, aligned with the moment in which the experimenter's fingers touch the object to grip it. The black dots indicate the beginning of each trial in which the object was presented to the monkey on a horizontal surface. There is no response from the neuron during this phase; it occurs during the grasping actions performed by the experimenter and by the monkey. Abscissa axis: Time; ordinate axis: The intensity with which the neuron fires, measured in spikes per bin = 20 ms.
Reprinted from *Brain*, 119 (2), Vittorio Gallese, Luciano Fadiga, Leonardo Fogassi, and Giacomo Rizzolatti, Action recognition in the premotor cortex, pp. 593–609, doi.org/10.1093/brain/119.2.593, Copyright © 1996, Oxford University Press.

This is necessarily a very brief overview of mirror neurons and their activity, as an exhaustive and technical disquisition on their multifarious properties in monkeys and the analogous mechanisms—sometimes even analogous individual neurons—found in the human brain would occupy too much space. There is a vast literature on the subject available to those readers who would like to add to their knowledge of the field.[40] However, keeping in mind the

---

[40] Rizzolatti et al. (2001, 2004); Rizzolatti and Sinigaglia (2006, 2010); Ammaniti and Gallese (2014). See Gallese et al. (2011) for a critical debate. For a radical criticism (which in our modest opinion is rather off the mark), see Hickok (2014).

main focus of this book, it is important to remember neuroscience shows that there is more to the sight of others performing an action than a simply pictorial reconstruction, according to an iconographic concept of sight. Watching someone performing an action is by no means the same as watching an apple fall; from a perceptive point of view, it requires we also simulate the action with our motor system, in our motor system.

The mirror neuron mechanism brings embodied simulation into the sphere of intersubjectivity and so is highly relevant for the understanding of all forms of mediated intersubjectivity that characterize narrative make-believe, as found in films.

Mirror neurons respond to the observation of an action being performed by others, even if the action is not completely visible, for example when the hand of the person performing the action and the object itself are obscured (Fig. 1.10).[41]

A part of the observer's motor system behaves as if the act were complete even if the final part—the achievement of the goal guaranteed by closing the fingers round the object—is invisible. Notwithstanding this, the motor mirror mechanism, the simulation of the observed action, still takes place. Fifty percent of the mirror neurons activate with the same intensity and the same temporal profile as when the action is visible. In our view, it is this activity that guarantees the continuity of the visual perception of the action in spite of the visual blackout. If we were to describe what we have just seen in cognitive terms, we would say that the observer has *inferred* the meaning of the observed action. This inference of the motor goal of third-party actions would appear to be mediated by the activity of the mirror neurons that control the implementation of the motor purposes of the same actions in the brain of the observer. The simulation reconstructs the part of the action that is not visible and therefore the goal of that action can be determined, but here we are talking about a motor inference, not a logical inference (Fig. 1.11).

It could be objected that observing an action by another person in which the final part is not visible and observing the same action from start to finish is quite a different experience. While this objection is more than justified it does not contradict our hypothesis; the same research study showed that the other 50 percent of the mirror neurons observed in the same conditions responded only when the complete action was visible. In other words, two different phenomenal experiences correlate with two different responses of the mirror mechanism. All the mirror neurons responded to the visible action;

[41] Umiltà et al. (2001).

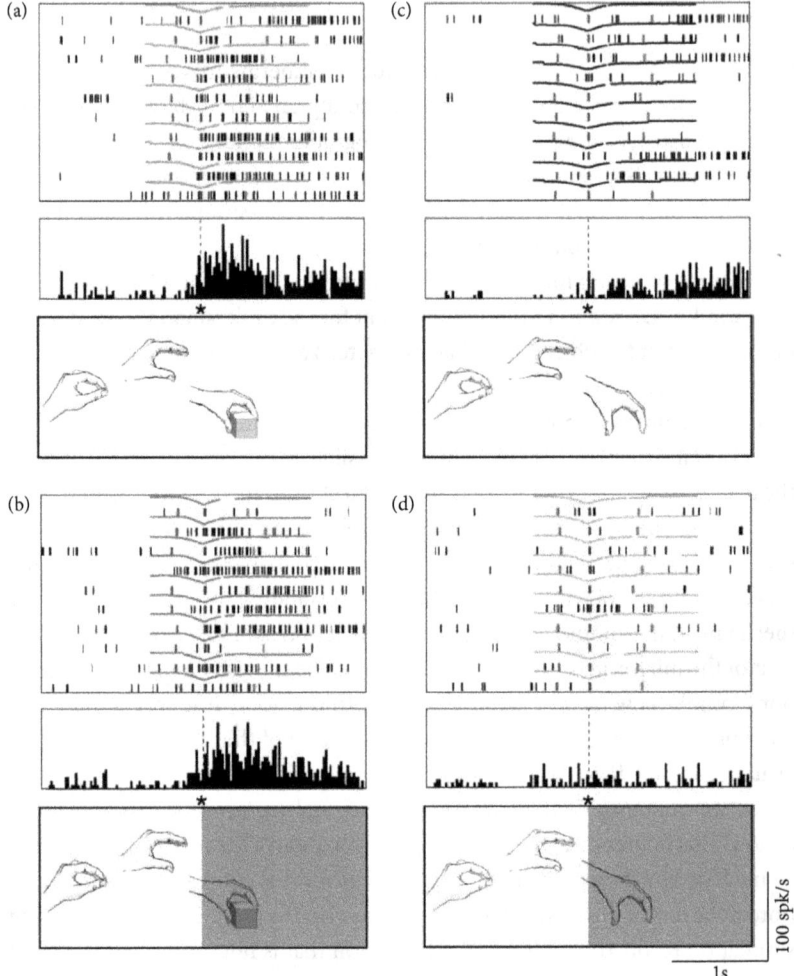

Fig. 1.10 Example of an F5 mirror neuron responding in conditions of full and partial visibility but not when the action is only mimed. The lower portion of each panel shows the action performed by the experimenter from the monkey's point of view. In (A) and (B) the experimenter moves his hand towards the object and grasps it. In (C) and (D) he mimes the act of grasping; the object is not present. The experimental paradigm consists of two basic situations: (A) and (C) full visibility; (B) and (D) partial visibility. In both (B) and (D) the area shaded in gray indicates the screen that prevented the monkey from seeing the final part of the action being performed by the experimenter. The upper portion of each panel shows the firing sequences and the neuron response histograms for 10 consecutive trials registered during the corresponding movements of the experimenter's hand. The vertical line indicates the point in which the experimenter's hand was closest to a photoelectric cell (in the partial visibility condition this corresponds to the moment when the experimenter started to move his hand behind the screen). Note that the neuronal activity in (A) and (B) is very similar but is almost absent in (C) and (D) (Umiltà et al., 2001).

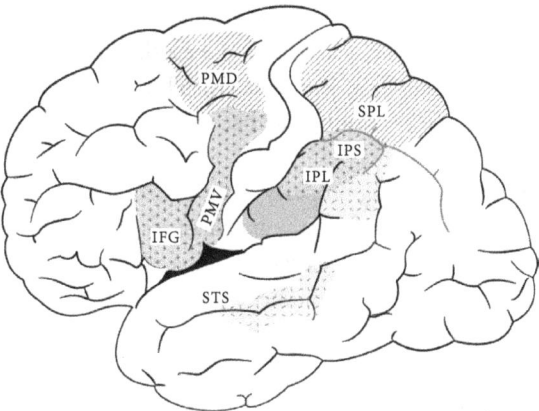

Fig. 1.11 The mirror mechanism relating to actions. A schematic representation of the cortical cerebral areas activated both while watching movements without meaning, motor acts performed with a specific goal in mind, and motor acts related to the use of objects.

PMD, dorsal premotor cortex; SPL, superior parietal lobule; IPS, intraparietal sulcus; IPL, inferior parietal lobule; PMV, ventral premotor area; STS, superior temporal sulcus; IFG, inferior frontal gyrus.

Reprinted from *JAMA Neurology*, 66 (5), Luigi Cattaneo and Giacomo Rizzolatti, The Mirror Neuron System, pp. 557–560, Figure 2, doi:10.1001/archneurol.2009.41 Copyright © 2009, American Medical Association.

only 50 percent responded to the version of the trial in which the attainment of the motor goal was obscured. We are aware that this does not justify stating without a shadow of a doubt that the diversified activation of the same neurophysiological mechanism *causes* the different phenomenal experiences of the various perceptive conditions of the same action. However, the fact remains that we find a sort of regularity from the personal experiential level to the underlying sub-personal level, as exemplified by the neural mechanism. Rules are very often generated by biological regularities.

Another property of the mirror neurons that is relevant for our work on cinema emerged from a research study carried out at Duke University.[42] Stephen Shepherd and his collaborators discovered that there is a class of neurons in the macaque's parietal lobe that controls eye movements in a certain direction and that also responds when observing another monkey looking in the same direction. This has revealed that the mirror mechanism is present also in eye movements, supplying the neurophysiological correlate to a well-known form of social behavior, the ability to share attention with others; when we see

[42] Shepherd et al. (2009).

someone looking in a certain direction or at a certain object, we direct our gaze in that direction or at that object. When we see a shot of an actor looking off-screen in a particular direction, immediately followed by a continuity shot of an object, we expect that continuity shot because we share the actor's attention, simulating his eye movement.

There is another interesting property of the mirror neurons that reveals the level of abstraction to which they are capable of responding, even in a prelinguistic species like the macaque. Certain actions are characteristically accompanied by sound. Imagine for a moment that you can hear footsteps in the corridor, followed by the sound of knuckles knocking on the door. These particular acoustic elements lead you to believe that someone has arrived and is waiting to come in; the noises help you to understand what is happening, even in the absence of any visual evidence. The sound of footsteps on the floor and of knuckles on the door leads us to imagine, in fact almost physically to visualize, the behavior of someone we cannot physically see. Just think how often these sounds have been used in cinema for diegetic purposes!

A series of experiments has been conducted to investigate the neural mechanisms linked to this ability. F5 mirror neurons not only respond when the monkey performs or observes an action, but also when it hears the sound that the action generates.[43] The sound of a nut being cracked by someone activates the same neuron that controls the cracking of the nut and responds to the sight of someone cracking a nut, even if there is no sound. Whether an action is performed or heard or seen makes no difference to these neurons. The activation of a part of the motor neural circuit that normally controls the execution of the action by the sound it produces, is yet another manifestation of embodied simulation of that action, obtained in this case by audio-motor multimodal integration. Analogous results have been obtained in studies on humans; in the human brain the sounds of actions activate a motor representation of those same actions in the person who hears the sounds. Neurological lesions that damage these premotor areas produce specific deficits; the motor deficit of a body part such as a hand or the mouth makes it more difficult to recognize an action performed with that same body part by another person, and this particularly so when only the sound is perceptible and the action is not visible.

We should add that the behavior of the visitor knocking at the door of whom we spoke earlier and which for the moment we have only described in terms of movement, actions, and sounds, is of course much more complex than this. The visitor certainly came with a precise goal in mind and to understand it we

---

[43] Kohler et al. (2002).

will need to combine a series of data and information: What we know about this person, the type of relationship that we have with him, our state of mind at the time, memories, and the possible association of preceding facts and events that have linked us to that person or to other parties who have relations with him ... the list is well-nigh inexhaustible. All these elements are necessary for a complete sense of the action, and certainly not all are "mapped" by the simulated motor mechanism triggered by the activation of the mirror neurons. What we are saying here is that in our opinion the simulation mechanism expressed by the mirror neurons is not potent enough to account for the multifarious psychological and cognitive levels that characterize every human relationship. To assume this would be madness, to confute it a waste of time.

The embodied simulation expressed by the motor mirror mechanism provides us with a notion of intersubjectivity connoted *in primis* and principally as intercorporeality, also at the sub-personal level of description typical of neuroscience. The sense of self comes into being precociously, starting from a principally physical sense of self and finds its origins in the possibility of interacting and acting with others. Daniel Stern's writings on this subject are well worth reading.[44] The simulation mechanism appears to be the neurological basis of how we tune into others, which is the point at which we start to build our sense of self and our understanding of others.

Our openness to the world and the intentionality of our mental processes, their being about something, are constituted and made possible by a primitive motor intentionality. The motor system, the repository for all our potentialities for a pragmatic relation with the world, prepares us to interact with it. Insofar as they are intentional, focused on the potential objects on which we intend to act, these motor potentialities define us as an intentional corporeal self. A pre-ordinated but flexible motor plan, of which the individual contingent relation modalities are coordinated every time, provides the supporting relational structure for self, the expression of which requires a close integration with the interoceptive and affective–emotional domain.[45]

Emotions, sensations, and embodied simulation

A few years after the discovery of the mirror mechanism for actions, it was thought that mirror neurons might be the tip of a much larger iceberg extending

---

[44] Stern (1985, 2010).
[45] See footnote 7.

to the domains of emotions and sensations, which until then had not been explored in relation to this neurophysiological mechanism.[46] The empirical evidence that subsequently came to light confirmed this hypothesis; other mirror mechanisms are, indeed, involved in our ability to share the emotions and sensations of others.[47] Of these mechanisms, the most explicit is that which regards facial mimicking; when we observe others expressing an emotion by facial mimicking, the facial muscles of the observer activate congruently, with an intensity that appears proportional to his empathic nature. This response is specific to the observations of facial expressions that convey emotions, as it is not evoked by the observation of facial movements with no emotional content.[48]

The observation and the production of facial expressions with emotional content could involve common neural structures, with functions hypothetically similar to those of the mirror mechanisms. In point of fact, both the observation and the mimicry of basic emotional facial expressions activate the same limited group of cerebral structures that include the ventral premotor cortex, the insula, and the amygdala.[49] The voluntary imitation of an expression of emotion, however, does not necessarily reproduce the subjective experience of the emotion that is being imitated. An fMRI study was first used to study this issue, examining cerebral activity in volunteers experiencing disgust provoked by inhaling unpleasant odors and while watching a video in which an actor expressed it with facial expressions (Fig. 1.12).

Seeing someone with a disgusted expression on their face activates the same portion of the left anterior insula that activates when we ourselves experience disgust.[50] This is even more interesting considering that a lesion of the anterior insula destroys both the possibility of experiencing disgust personally and of recognizing this emotion in others.[51] When we see a facial expression, we do not understand the meaning behind it simply by explicit inference by analogy. The emotion being expressed by the other person is first and foremost constituted and directly understood by re-using the same neural circuits on which our first-hand experience of that emotion is based. When we observe the facial expression of an emotion we also simulate it internally. This has been further confirmed by clinical neuropsychology; the integrity of the sensory–motor system is essential for recognizing the emotions externalized by others,[52]

---

[46] Goldman and Gallese (2000); Gallese (2003).
[47] For a recent review see Ammaniti and Gallese (2014).
[48] Sestito et al. (2013).
[49] Carr et al. (2003).
[50] Wicker et al. (2003).
[51] Adolphs et al. (2003); Calder et al. (2000).
[52] Adolphs et al. (2000).

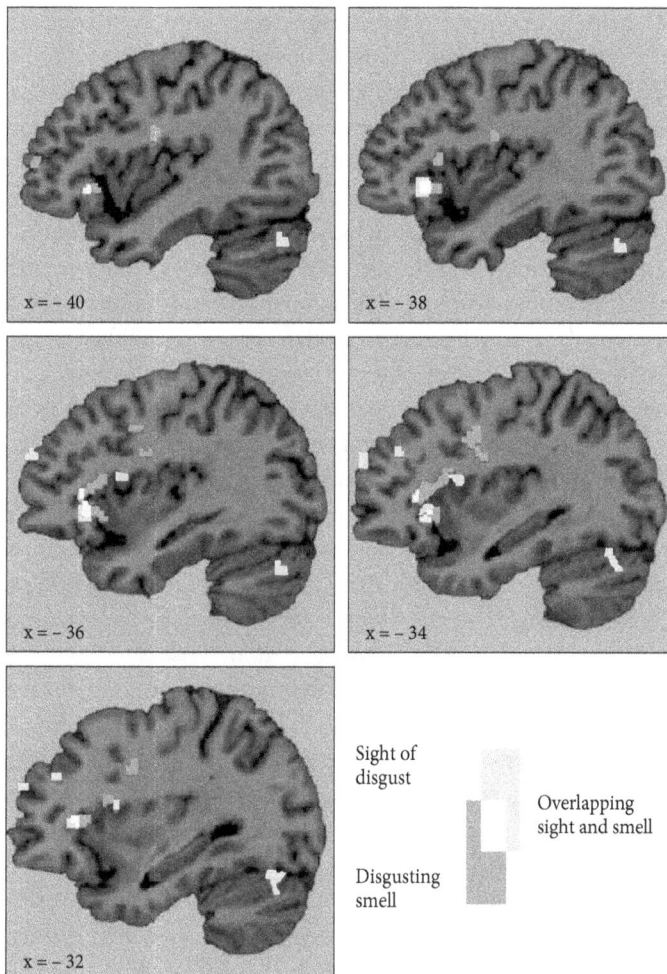

Fig. 1.12 Activations caused by unpleasant olfactory stimuli and by facial expressions evincing disgust. The light-gray areas indicate the activations following the viewing of the facial expressions; the activations stimulated by the unpleasant odors are shaded in dark gray; the white areas are those activated by both stimuli. The activations are shown in parasagittal sections of a standard Montreal Neurological Institute brain (Wicker et al., 2003).

because it helps to reconstruct what we would feel in any given emotional situation, through simulation of the relative corporeal state.

However, from an experiential point of view feeling an emotion and observing the facial expressions of others are two very different conditions. When we see an expression of disgust forming on someone's face, we recognize the emotion without necessarily having to feel it ourselves. The results of the experiment described were even more intriguing as the stimuli used were short films in which an actor interpreted disgust, expressing the emotion through characteristic grimaces. It is important to keep in mind that the conclusions drawn from this study have been used to explain the cerebral mechanisms that control our recognition of disgust being experienced by others in real life, while in this book we have to deal with another problem: How our brain–body manages to steer us through the passage leading from the real world to the world of fiction. Watching a film that lasts no more than a couple of seconds during an fMRI study is certainly not the same thing as watching a real full-length movie, comfortably ensconced in a cinema seat. All the same, the fictional base of both types of image remains.

It is extremely rare for real-life films to be used in neuroscientific studies on the brain for reasons of standardization and practicality; usually, studio reconstructions of real-life situations are used. In those rare cases where individual neurons of the macaque[53] or an entire human brain[54] have been studied in comparative experimental situations using a human being in flesh and bone, it has been seen that the neural or cerebral response differed in intensity: On average, and particularly in humans, the physical presence of another living being evoked a more intense response than an apparition on the screen.

We do not believe, however, that the brain differentiates the real or fictional nature of an image or scene solely on the grounds of a different intensity of the same response. An fMRI study compared three different situations focused on the experience of disgust: (1) the subjective experience of imbibing a liquid with an unpleasant taste; (2) observing expressions of disgust on the faces of actors; (3) readings of short stories generating disgust.[55] The results show that all three conditions activated the left anterior insula, the very same region that was activated by Bruno Wicker and his collaborators in the experiment described earlier. That said, this region was functionally connected to different cerebral areas and regions in the three experimental conditions; in the first

---

[53] Ferrari et al. (2003); Caggiano et al. (2011).
[54] Järveläinen et al. (2001).
[55] Jabbi et al. (2008).

condition the activation of the left anterior insula most likely generates the genuine experience of disgust, while in the other two conditions it is simulated, evoking the same content. Its connections with the diverse cerebral circuits likely account for the different subjective experiences characterizing the three conditions.

These results and, in more general terms, the approach and thesis expounded in this book, show how neuroscience can make a concrete contribution to long-standing philosophical and aesthetic debates on the nature of emotions evoked by fiction. Are the emotions that we feel when we watch a movie real, analogous to those we experience in similar situations in real life? How is it possible that we feel real emotions for situations that we know for certain are not "real"? Unfortunately, we do not have enough space for a full-scale critical discussion of the literature on the theme, so we will limit ourselves to proposing that the theory of embodied simulation can contribute to explain both the genuine nature of the emotions generated by fiction and the various reactions it provokes in the audience. The fear, anger, and disgust that we feel when watching a movie will never generate the same behavior that the vision of similar situations in the real world would normally provoke in us. The combination of the simulation of the somatic and visceral–motor content of a given emotion (such as that produced by the activation of the anterior insula for disgust), with the activation of the cerebral circuits that regulate and control different behaviors and strategies, would appear to provide an interesting solution to these problems, or at the very least a new approach and basis for discussion. If we add to this the cognitive factors of *framing* that characterize the experience of recounting fiction, we are coming even closer to understanding why it is not so paradoxical to feel real emotions for situations that are obviously fictional, situations that have been created by a filmmaker or scriptwriter.

Similar simulation mechanisms have also been described in connection with the perception of pain[56] and touch. When we see the body of another person being touched, caressed, slapped, or wounded, it activates those parts of our motor, somatic sensory, and visceral motor/limbic systems that normally

---

[56] Botvinick et al. (2005); Jackson et al. (2005); Singer et al. (2004). On the question of pain, Hutchison et al. (1999) reported that during a neurosurgical operation a single neuron was seen to activate not only when a painful stimulus was applied to the patient's body, but also when he saw the same stimulus being applied to the body of the neurosurgeon. This neuron, recorded in the anterior cingulate cortex, one of the cerebral areas together with the anterior insula that activates for first-hand experience of pain and when pain is observed in others, seems to express mirror properties for this sensorial modality. In our opinion, combined with the information obtained from fMRI studies, this fact makes the theory of embodied simulation in this domain even more plausible. A recent study by Carrillo et al. (2019) showed evidence of mirror neurons for pain in rats.

guide our behavior and map the sensory motor, tactile, nociceptive, and interoceptive sensations that we personally experience.

Our research group was the first to show that one of the cerebral areas that maps our tactile sensations is also activated by the tactile experiences of others with the same body part (Fig. 1.13).[57] Sharing tactile experiences will be discussed in greater depth in Chapter 5.

The emotions and feelings experienced by others, independently of whether they are real or not, are first of all constituted and directly understood through a reusing of a part of the same neural circuits on which our first-hand experience of these same emotions and feelings are based. Our involvement with the situations and emotions of the characters in a movie is probably mediated, at least in part, by this mechanism.

The next section examines how our brain–body system permits us to pass nonchalantly from reality to fiction and vice versa.

The person as a corporeal form between the real world and the world of fiction: Liberated simulation

Being a person does not just mean experiencing physical reality, it goes farther than that. It encompasses the ability to conceive other worlds, to let the imagination wander, and only recently in the history of our species, relate to a reality that is not immediately definable as real or fake, depending on the means by which it is perceived. Ever since Orson Welles' radio announcement that aliens had landed on Earth, the fact that an event is reported on the media is no longer a guarantee of its truth. We have at least two visions of the world: "*Zuhanden,*" within reach, during our daily round in our body in the physical world; but also "*vorhanden,*" presence at hand, in that real events are externalized through narration (originally by storytellers, the oral tradition, now on newspaper articles, news reels, documentaries) and as fictitious renderings *of* and *on* real events (novels, films, television series).

We are convinced that with the support of contemporary neuroscience we are in a position to infer *Vorhandenheit* from *Zuhandenheit*, basing this inference on the hypothesis of a progressive mediated and artificial externalization of the representational corporeal formats that originally evolved to facilitate our contacts with the real physical world. Our hypothesis is that the abstract externalization derived from the representation of what is real,

---

[57] Keysers et al. (2004).

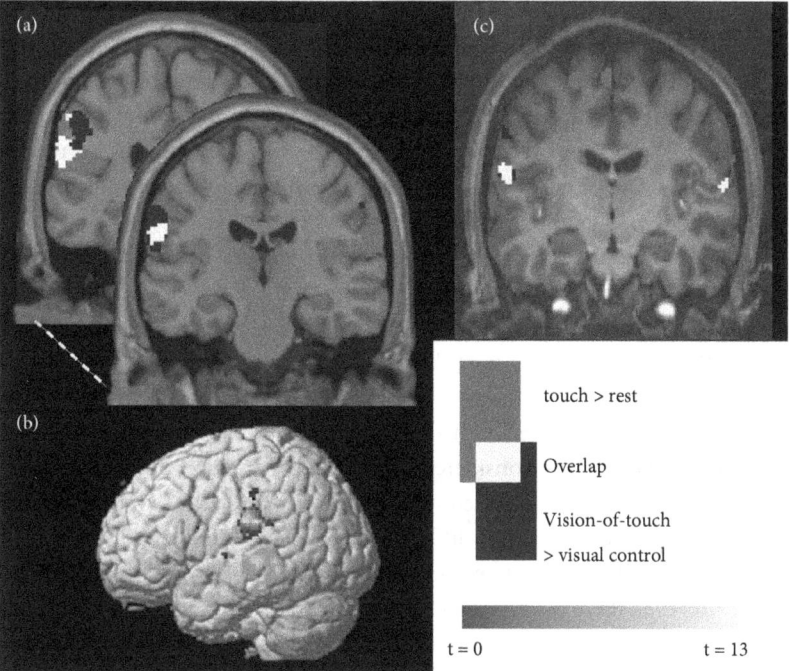

Fig. 1.13 Statistical maps showing the cerebral regions activated by a first-hand experience of feeling a touch on the legs and by observing tactile stimulation of someone else's legs ($p > 0.001$).
(A) Two coronal sections extracted at Y = −30 and −20 show the extension of the cortical region activated by the first-hand experience of touch, and by observing tactile stimulation of touch applied to another person. The regions activated by the sensation of a touch to the legs are indicated in dark gray; those activated by observing tactile stimuli applied to the legs of another person are indicated in light gray; those that activate in both conditions are indicated in white. (B) Illustration of the lateral surface of the brain regions activated in both conditions. The gray scale (bottom left bar) indicates the intensity of the activation shared in both conditions. (C) Illustration of the activations recorded in a single subject at Y = −16 (Keysers et al., 2004).

and first and foremost formulated by language, has *its roots in transcending the body while remaining within its bounds*, according to the principle of embodied simulation.

Neuroscience has helped us to understand that the boundary between what we call the real world and the imaginary and imagined world is far less clear cut than might be thought. As we have just seen in the preceding paragraph, experiencing an emotion and imagining it are based on the activation of cerebral circuits that are identical in parts. Analogously, seeing and imagining that we are seeing,[58] doing something and imagining that we are doing something,[59] share the activation of parts of cerebral circuits they use in common. This dual activation is a further indication of embodied simulation and the correlated notion of neuronal "reuse." This case also shows how embodied simulation exceeds the notion of empathy.

We can hypothesize that the imaginary worlds we create and inhabit are possible due to the embodied simulation mechanism combined with the ability for cognitive abstraction. A recent research study using high-density EEG has shown that the cerebral circuits that inhibit an action being performed and those that actually block the performance when we really only want to imagine it are partly the same.[60] Our ability to conceive imaginary and sometimes non-existent worlds was considered by many and for many years as an exclusive prerogative, with high added value, of the human species. But are we sure that this is really the case? It is difficult to attribute to other species the ability to represent to themselves that they want what they want or what they might want. However, there is a certain level at which it is possible that other animals might represent and imagine what they want. Taking a long jump back in time, it is worth consulting the reflections of Aristotle on this subject.[61] In Aristotle's view, animality is essentially characterized by two aspects: Sensation/perception (*aesthesis*) and desire (*orexis*). According to Franco Lo Piparo's interpretation,[62] these two aspects are closely intertwined. In fact, perception cannot be considered as a mere registration of what is present in the world. Lo Piparo says that that is what machines do. The simple recording of information is what characterizes a computational mind. In animals, however, perception is always accompanied by feelings of pain or pleasure, and most of our own behavior can be explained in terms of this intertwining of perception, pleasure, and pain. Aristotle, and many who came after him, including Sigmund Freud, held that we are stimulated to act by the desire for pleasure and the need to keep pain at bay; we want to have what causes us pleasure and avoid anything that causes unpleasantness or pain, and act accordingly. Desire, perception, and the pleasure/pain duo are crucial aspects in determining our behavior. According

---

[58] Farah (1989, 2000); Kosslyn et al. (1993); Le Bihan et al. (1993); Kosslyn (1994).
[59] Roland et al. (1980); Fox et al. (1987); Decety et al. (1990); Jeannerod (1994); Parsons et al. (1995); Porro et al. (1996); Schnitzler et al. (1997).
[60] Angelini et al. (2015).
[61] Gallese and Cuccio (2015).
[62] Lo Piparo (2003).

to Aristotle's line of thought, the desire that drives us to act implicates the notion of mental representation. No animal can desire something of which it has no representation, unless of course that something is (for example) directly present in its immediate perceptual context. That said, not all wishes and desires are present in the immediate context; animals often have to look for what they want, whether it be food, water, or a mate. The search for something that is not immediately present implies a representation of that something. Its absence, that which is not or which is not yet, becomes the paradigm of the representation. Aristotle defined this level of representation, deeply rooted in our senses, as *phantasia sensibile*, "sensitive phantasy" (Aisthetiché). In fact, being endowed with sensitive phantasy means to live a certain type of experience in the absence of what normally triggers the experience. This Aristotelian concept of *phantasia sensibile* brings to mind the phenomenon of mental imagery we talked about earlier. Imagining or representing something really does activate the areas of our brain linked to the related experience. Humans share this type of representation with other animals. It is very difficult to resist the temptation to interpret the results of the experiment conducted by Maria Alessandra Umiltà and her collaborators,[63] in which half of the mirror neurons being studied continued to respond, even when the hand that was grasping the object was no longer visible, as the neural correlate of the monkey's imagination. Humans, however, are endowed with *phantasia linguistica* (logistiché), as well as *phantasia sensibile*. Language provides us with a new modality of representation that does not exclude our motor and sensory corporeal knowledge, and indeed uses it in its original format. In our view, the dynamic blending of the corporeal and symbolic dimensions produces human specificity and gives form to *liberated embodied simulation*. Together with symbolic representations, liberated embodied simulation enables us to build other possible, parallel, and imaginary worlds.

We human beings constantly inhabit a plethora of real and imaginary worlds, or, at least, we have the ability to do so at will. This depends on the fact that our relation with reality, even when the only representation is ours, made possible by our sensory motor processes, is always "virtual" to the extent that it is subjective, filtered by our previous experiences and, most important of all, the final product of a constant sociolinguistic negotiation and construction. That said, we are convinced that the difference between reality as we see it with our own eyes and reality mediated by theater, cinema, or other forms of dramatic art is dimensional rather than categorical.

[63] Umiltà et al. (2001).

We will discuss liberated simulation as a characteristic trait of our lived experience, also and above all when we are watching a movie. When we open our minds to aesthetic experience as we do when we go to the movies, we are forced to temporarily suspend our grip on the world, freeing up energy that previously we were not able to access, and putting it at the service of a new regional ontology that will show us new aspects of the world and of ourselves.

From a cognitive point of view, when we watch a movie and lose ourselves in its plot, we temporarily suspend incredulity, refrain from judging what we see to be unreal, pretend that the movie we are watching is not a fabrication of reality. We feel that stopping at this point means endorsing a partial image of the real experience of watching a movie. Attributing the experience and the understanding of the narrative fiction of the movie exclusively to the conflict between a series of beliefs regarding the world means to amputate what, in our opinion, is an essential component of the spectator's movie experience; the corporeal experience on various levels due to the mechanisms connected to the perception of the real world.[64] Rather than a suspension of disbelief, the aesthetic experience generated by watching a movie can be described as "liberated simulation," as the outcome of a strengthening of the mirror and simulation mechanisms described earlier. When watching a work of fantasy, our relationship with it is free of the normal, direct personal involvement with daily reality. We are free to love, hate, be terrified, be delighted, all from a safe distance. It has often been pointed out how advantageous it would be from the adaptive point of view if we were able to play out and simulate everyday situations through fiction or by participating in ancient drama forms. It is this safe distance between ourselves and what is being played out on the screen that we find so pleasurable, and makes the mimesis cathartic, bringing fully into play our natural mimetic openness to the world.

With embodied simulation, as in real life, we experience a decoupling of the effective action that, however, is even more radical. When watching a movie, or performing any form of activity connected with art, such as reading a novel, watching a play in a theatre, observing a painting or a sculpture, we are usually immobile. We let our guard down, just temporarily; we are no longer in the constant state of alert and readiness to anticipate an event and respond adequately that characterizes our daily engagement with the real world.

---

[64] However, certain writers maintain that incredulity should be applied and used while watching movies in order to balance the effect of realism that we perceive through our senses. See Barratt (2007, pp. 63–4).

When we settle into our seat at the cinema, we don't have to escape, defend ourselves, or fight. Our perception of what is happening on the screen could almost be said to be neotenic in nature;[65] in other words, similar to our perception in that early period of development when, due to our scarce motor autonomy, our interaction with the world is mediated principally by a simulated perception of what goes on around us in terms of events, actions, and emotions. It is very likely that we learn to calibrate gestures and expressions and to associate them with pleasant and unpleasant experiences by observing others, through embodied simulation and its flexibility.

When we are at the cinema, our state of relative immobility and disconnection from the world is intentional and not due to the immaturity of our sensory–motor development. All the same, this immobility, caused by a greater degree of motor inhibition than usual, probably allows us to reallocate our neural resources, intensifying this type of non-linguistic representation and ensuring a more intense adherence to what we are simulating. As Edgar Morin wrote, "when the influences of the shadow and the double merge on a white screen in a dark room and the channels of action are blocked, the locks of the myth, the dreams, the magic are opened for the spectators, sunk in their alveolus, an entity closed to all except the screen, wrapped in the double placenta of an anonymous community and darkness."[66]

Seated in the cinema, our interaction with the world is exclusively mediated by a simulated perception of events, actions, and emotions represented in the movie we are viewing. It is almost as if there were a form of emotional *transfer* between actors and spectators; the latter, being forced into inactivity, are more receptive to feelings and emotions.[67] When we watch a movie, not only do we focus our attention exclusively on the screen, but our immobility releases all our resources of embodied simulation and uses them to create an absorbing relationship with the characters of the plot.[68] This is an effective way of reflecting on the differences comparing the mental states that depend on our

---

[65] Neoteny, literally "tending towards youth" (νέος τείνειν), is the term used to indicate a retarded and/or prolonged physiological or somatic development of an organism. It is particularly relevant for the human species, which is characterized by the fact that the maturation of the central nervous system continues for many years after birth. In our view, this aspect, frequently overlooked when explaining the apparent uniqueness of the human species in the animal kingdom, has a certain importance. Our neotenic nature shows the extent to which the development of our brain takes place in a social environment and is therefore extremely conditioned by the quantity and quality of the social relations we manage to establish with others, starting with our parents. The neoteny of the human species is another biological argument in support of the crucial importance of society and relationships for the development of our cognitive faculties.
[66] Morin (1956, p. 106).
[67] Ortoleva (2013).
[68] Wojciehowski and Gallese (2011).

aesthetic attitude and those related to our everyday consciousness and their differences.[69]

After this exposition of our theory of the relationship between cinema and neuroscience and its focal point, the embodied simulation theory as a model of perception, this last section of the first chapter provides a review of what has been discussed up to now on various levels and with different objectives, regarding the encounter between neuroscience and cinema.

Brain–body and cinema

For the entire duration of the twentieth century cinema represented, and to a certain extent still represents today, the most extraordinary and realistic reproduction of our life. Everything, from video games to virtual reality, that has to do with virtual environments and images in movement, has taken the intersubjectivity and agency models experimented and characterized in cinema as best practices. Every movie is essentially based on the actions and interactions of characters who move in front of our eyes just like the people we see every day in the street; those forms on the screen make the same gestures, feel the same sentiments and emotions, and are driven by forms of intentionality, all of which faithfully reproduce the workings of the real world. To obtain this impression of continuity with reality and create the conditions for highly efficacious forms of mediated intersubjectivity, a movie has to reassemble, reinforce, and in certain cases modulate *ex novo* all those aspects of reality necessary to hold the spectators' attention and facilitate their emotional involvement. For a movie to be understood and possibly to be a box-office success, the plot must be clearly oriented to a well-defined objective, the feelings, sentiments, and intentions of the characters must be clearly and precisely drawn, and, contrary to what happens in everyday life, it must have a certain logic and be meaningful and accessible from both a perceptive and cognitive standpoint.

As we will see later on, filmmakers have various ways of creating the conditions necessary to reinforce and dramaturgically modulate the essential components of cinema: Movement, action, interaction, gestures, sentiments, and emotions, which unfold in a bi-dimensional space that gives the illusion of being three-dimensional and in which we feel we know our way around, even to move around. Our brain–body system not only gives us experience of that

[69] Smith (2009, p. 44).

virtual space and allows us to process a spatial cognition that we need to live a given experience, it also puts us in a position to share the situations, actions, gestures, and emotions that take place in that other dimension represented on the cinema screen.

In a book written with the objective of helping a vast public to understand our psychophysical reactions to films, Jeffrey Zacks maintains that the way our brain reacts to real situations, those situations for which it evolved, is also the way it shapes our reactions to what we see in the movies. Zacks condenses this approach to the real world and the world of film in two rules: The "mirror rule" and the "success rule."[70] The first, the mirror rule, is "do what you see," while the second, the success rule, is "do what has worked." In other words, cinema is the art form that more than any other brings us face to face with situations similar to those of the real world. Certain situations we reproduce physically, internally as adults or externally as children often do, jumping up and down when watching an action movie; others require rapid decisions, dictated by precise stimuli and not limited to motor responses, and refer to more complex forms of social interaction that are capable of modifying our behavior.

Zacks makes sure his readers are aware that there are no centers in the brain delegated to process his mirror or success rules, but that various cerebral areas share the task of enabling our experience as spectators. Filmmakers have had to learn very quickly how to make films with high impact on the senses and minds of the spectators, films that boost the processes of simulation and inherence starting from the functioning of the brain. We do not see anything reductionist in stating, as Joseph Anderson did, that a movie is a program that can be read by one single processor, the human brain[71] (this hypothesis was reinforced by the outcome of the experiments conducted by Asif Ghazanfar et al. on monkeys at the cinema),[72] nor as Antonio Damasio wrote, that those who created and rapidly codified a shared movie style have tested its workings (certainly using instinctive and inductive processes) on certain characteristics of the human brain.[73] In an essay on the forms of stimulation of the so-called working memory by films, Yadin Dudai wrote that, given the rapidity with which cinema had risen to the fore among the various disciplines of modern art and culture, it could be said that "the human mind has adapted to cinema so rapidly and successfully because the human brain has a neural

---

[70] Zacks (2015, pp. 4–11).
[71] Anderson (1996, p. 12).
[72] Ghazanfar and Shepherd (2011).
[73] Damasio (2008).

system, originally evolved for other purposes, which almost called for cinema to be invented once the technical elements became available."[74]

These statements reflect the most recent trends, undoubtedly pervasively and strongly driven by the new brain-imaging techniques, in the dialogue between cinema, physiology, and the brain sciences; Italy has been one of the most precocious centers of this debate since the beginning of the twentieth century.[75] Later we will see how physiologists, psychologists, philosophers, and filmmakers interrogated spectators' brains and bodies in an attempt to assess our predisposition to images in movement on the one hand, and the peculiarities and future prospects of this new mass phenomenon, on the other. In 1986, after publishing important works on cinema, Gilles Deleuze went so far as to say that linguistics and psychoanalysis have very little more to say about film, while interesting novelties on the subject could come from brain biology.[76]

On the cinematographic side of the field, neuroscience benefitted from a composite debate, which had its origins principally in North America from the end of the 1980s through the 1990s, following on the crisis of the so-called Grand Theory, mainly inspired by the semiotic and structuralist paradigms, and in certain cases based on Lacanian psychoanalysis. In the 1990s, the decade proclaimed by the late George H.W. Bush as the Decade of the Brain, interest in phenomenology of cinema sprang to life with the contributions of Allan Casebier and Vivian Sobchack; their books on the subject, with interpretations from Husserlian and Merleau-Pontian perspectives, respectively, were published almost contemporaneously.[77] The Merleau-Pontian perspectives were particularly successful, as Sobchack was skillful in not completely decoupling them from an "embodied" resolution of certain semiological strongholds and, above all, from the analytic affinity to films. The tradition of phenomenological studies, which had given a decided boost to the increasing interest in tactility and haptic visuality, strangely enough managed to avoid excessive contamination by neuroscientific literature, particularly from the start of the last century, in spite of the fact that Merleau-Ponty became the preferred source of material for neuroscientific papers with a philosophical and artistic bent.

As was to be expected, the other important model used by neuroscience was constituted by cognitive studies on film, whose contraposition with the paradigms of the Grand Theory became the manifesto of an influential school of thought, whose principles were expressed in an important volume

---

[74] Dudai (2008, p. 22).
[75] See Alovisio (2013).
[76] Deleuze (1986).
[77] Casebier (1991); Sobchack (1991).

entitled *Post Theory: Reconstructing Film Studies*, edited by David Bordwell and Noel Carroll.[78] While phenomenology offered decisive starting points as far as the body (of the spectator and of the film) was concerned, the cognitivists had concentrated on the mind and had immediately placed their bets on what they defined as a "naturalistic turning point,"[79] cross-checking the cognitive theories with data that came mainly from the psychology of vision. The contribution of neuroscience still hasn't been completely assimilated in the cognitivist debate, possibly because it has tended to side with cognitive (and corporeal) archeology and on more than one occasion has aimed at restructuring certain positions, in particular those linked to simulation and identification processes. In the last few years, there have been opportunities for exchanges with the neuroscientific community, particularly at the Society for Cognitive Studies of the Moving Image (SCSMI) thanks to the commitment of psychologists, psychologists of vision, and neurophysiologists who have started to work on cinema, as have James Cutting, Jeffrey Zacks, Joseph Magliano, and Tim Smith, as well as those who only make contributions every so often but always with great impact, above all Uri Hasson and Art Shimamura.

Before the *viewer-as-body* of the phenomenologists and the *viewer-as-mind* of the cognitivists (or even at the intersection of these two models), according to Adriano D'Aloia and Ruggero Eugeni[80] there was the *viewer-as-organism*, a natural subject for neuroscience. While studying this new viewer model and so gaining an understanding of the attitudes of the *viewer-as-body* and the *viewer-as-mind* during analysis, neuroscience not only has the possibility of problematizing the strong models of cinematographic spectatorship (of which we will say more later), but also of testing the validity of the approach to cinema proposed around the middle of the 1900s by French *filmologie*, which included integrating various fields of knowledge (physiology, psychology, psychoanalysis, anthropology) into the theoretical study of cinema. From this point of view, D'Aloia and Eugeni chose to coin the neologism "neurofilmology," which as opposed to other neologisms proposed by, for example, Hasson ("neurocinematics"), Shimamura ("psychocinematics"), or Ghazanfar ("evolutionary cinematics"), suggests a greater degree of attention to the theoretical/epistemological contribution of the neuroscientific approach.

---

[78] Bordwell and Carroll (1996).
[79] Bordwell (2013).
[80] D'Aloia, Eugeni (2015, pp. 19–20).

This three-pronged approach to the viewer/spectator has the advantage of helping us to understand clearly the level of the contribution of neuroscience and the extent of the range of its influence. At the same time, indirectly, it obliges us to face the wide range of uses of neuroscientific models (which combine many theoretic and philosophical reflections from over a century of thought about cinema) and the high coefficient of speculation guaranteed by evidence that emerged from experiments that were not intended as studies of cinematographic experience. While "neurocinematics" and "psychocinematics" are more experimental in nature, with their two-fold objective of discovering something new regarding both the human brain and cinema, "neurofilmology" is more theoretical, waiting to judge the reach of neuroscientific findings on cinema (but not only) and looking for the essence of these findings in a critical comparison between the experimental slant and the theoretical framework. In this context the "neuro-" prefix does not immediately signify experimental data (as in the case of Hasson), but can be interpreted in the wider and figurative sense, a new season for the theory, the observation of the cinematographic phenomenon at close quarters,[81] or a new filmic form.[82]

Our position oscillates in what we hope is a balanced approach between these two options of theoretical vitality and experimental necessity, which must be the objective and the testing ground for experimental aesthetics, in general, and the study of film, in particular. As already mentioned in the Introduction, this dual approach is greatly facilitated by the fact that we come from two very different walks of life, each with our own specific scientific training, and the shared belief that every discussion and every experiment has something new to contribute to the knowledge of cinema and the ways in which our brain–body system responds to forms of mediated experience that are increasingly a part of our daily existence. We have developed a series of competences regarding these forms in real life and in the assimilation of codes and meanings for which it is still necessary, as Mark Johnson would say, to return to the corporeal roots from which they originate.[83]

This book is not a general work on cinema and neuroscience, studying psychophysical reactions to films. Based as it is on the theory of embodied simulation, it examines a precise aspect of the embodied relationship with film techniques and, consequently, with the fictional horizons they delineate. We are convinced that the theory of embodied simulation has the potential

---

[81] Guerra (2015a).
[82] See, for example, Pisters' reflections (2012).
[83] Johnson (2007).

to reject all accusations of reductionism; in fact, it is not sufficient to limit the risk of an excessive naturalization of the movie experience balancing neuroscientific data with theoretical sources, whose function is to act as a control system and center for hypotheses, as well as a generator of questions and answers that can benefit from the neuroscientific contribution. A process of investigation has to be identified that is both open to dialogue and contains the seeds of a debate encompassing both the aesthetic element and the models of analyses and studies that have acted and continue to act on it. From this point of view, embodied simulation as described in the preceding paragraphs seems to be the best way of communicating and relaunching the idea of "neurocinematography" on a theoretical and experimental basis that reveals its full potential, but at the same time does not elude criticism.

Cinema is a vision of the world in all its composite immensity and fascinating elusiveness, a vision of the world that widens our perceptual and cognitive horizons and exercises our imagination. It plummets into the depths of a dialogue with its times and history, providing us with an incredibly complete interpretation and modeling, ranging from our relationship with technology to our social, cultural, and political life.

When we surrender ourselves to the pleasures of the movies, it is difficult not to be amazed by the power cinema has to sweep us into another, virtual dimension in which we always find a sense of the real world we have just left behind. As the philosopher Jacques Rancière wrote, cinema is a multiplicity that tends to reject uniform theories: It is a place where we allow ourselves to be enthralled by shadows on a screen, it is the memory and the story we retain of those shadows, and the more it enters into our personal experiences, the more it deviates from them; it is an "ideological machine" that injects images into society, images of the past, present and the future, it is an art form that pulls apart the very concept of art, it is a philosophical concept which as it develops tends to break free paradoxically from its essence of cinema but genuinely becomes philosophy.[84] Many critics of the theory of cinema are afraid of losing this multiplicity, of having to sacrifice the totality of the experience, which, in fact, owing to its very multiplicity, requires diverse inter-communicating methods of approach in order to have an exhaustive concept of the presence of cinema in our lives.

It is by no means feasible to assume that cognitive neuroscience alone can account for the complexity of the filmic experience or the modalities of the relationships between humans and the possible worlds of art. Reductionism, which

---

[84] Rancière (2011, pp. 28-9).

continues to be the great bugbear in the processes of contamination between the humanities and scientific studies—we prefer these terms that are slightly less controversial than "human sciences" and "exact science"—is curbed from the start when neuroscience engages in dialogue regarding a critical theory of our life in the world (of which fiction, in whatever form, is undoubtedly a part). We are convinced that the recent literature regarding embodied simulation (in fields as far apart as neuroscience, philosophy, art, cinema, and theatre) has brought to light two relevant points: Firstly that the study of motor and non-propositional components of our experience, both real and mediated, is decisive in understanding the overall picture of the intrinsic pragmatic and performative nature of our intentional relationship with the world around us (in which artistic objects, as such, in turn connote a reference to our presence in the world); secondly, for a discussion on the subject to be complete and rational, able to provide answers that are only apparently banal to questions on our desire for make-believe (and in this specific case, regarding our vision of the world as seen on the cinema screen), there must be dialogue between the sub-personal, functional, and phenomenological levels—in other words starting from the neural circuits and arriving at intersubjectivity.

It is no longer a question of whether neuroscience should get involved in the study of art, philosophy, or economy, ethics, and law; it is a question of why it gets involved and what form the involvement should take. Once again, film provides a good testing ground. Many neuroscientific experiments use stimuli in video form, moving images of individuals carrying out particular actions to trigger reactions in the participants and which therefore imply a mediated response (just as an example, as we will see later, Shimamura maintained that very often films with fictional content offer more interesting and functional dynamic stimuli than those created ad hoc for the purpose). Both in these videos and in other cases (the works of Hasson, for example)[85] where films with fictional content were in fact used, the objective is to use the film simply as a stimulus to generate results for studying memory, sexual arousal, facial recognition, or conscious experience.[86] In certain cases, as in the famous experiment conducted by Hasson et al. that launched the new discipline known as "neurocinematics," the question of the control that a filmmaker can exert over the mind of the spectator by adopting particular stylistic techniques and the willingness of the spectator to "let himself be controlled" in order to fully enjoy

---

[85] Hasson et al. (2004, 2008).
[86] Of the many studies available, see Bocher et al. (2001, pp. 105–17); Iwase et al. (2002); Rothstein et al. (2005); Furman et al. (2007); Nishimoto et al. (2011); and Naci et al. (2014).

the physical and mental involvement exerted by the movie, has undoubtedly given rise to issues that to a certain extent could have been foreseen, but have never been so clearly formulated.

An example of direct cinematographic inherence can be found in a recent fMRI study[87] that examined the empathic components of film reception, studying the cerebral circuits activated by the viewing of sequences with high emotional content (the imminent loss of a loved one) taken from two famous movies (*Sophie's Choice* by Alan J. Pakula and *Stepmom* by Chris Columbus). The neuroscientists applied an approach based on the concept of triangulation[88] between phenomenological aspects, a psychological description and neural correlates, consisting of the evaluation of the network coherency index—a parameter that measures the extent to which cerebral areas are dynamically interconnected during the elaboration of various stimuli (in this specific case, the various sequences of the movie). The results of the study showed the activation of two brain circuits: The first constituted by the anterior cingulate cortex and the anterior insula, described as the embodied simulation circuit for pain, whereas the second circuit was constituted of prefrontal and temporo-parietal cerebral areas that are usually described as the circuit for the Theory of Mind, the activation of which is normally correlated with explicit mentalization tasks. The authors of the study interpreted these results as evidence of the presence of distinct cerebral circuits at the base of the, respectively, implicit and explicit components of empathic involvement.

The experiments on editing carried out by Jeffrey Zacks and Joseph Magliano in light of theories on the so-called event segmentation,[89] are another example of direct cinematographic inherence. On the one hand, their work provides a new slant on continuity editing (to be compared with the eye-tracking studies on the attentional theory of continuity editing by Tim Smith,[90] to which we will add our high-density EEG studies in Chapter 4) and, on the other, it shows that it is no longer feasible to talk of top-down segmentation of the events in a movie; the time has come to start thinking in terms of bottom-up mechanisms as prompted by the physical elements that mark the transition between the shots and facilitate forms of embodied relations with the subject of the movie.

The works cited here and a number of studies we ourselves have conducted justify stating that sensory–motor and affective contents should be included in the elements used to study the cognition of cinematographic narrative.

[87] Raz et al. (2014); see also Raz and Hendler (2014).
[88] See footnote 24.
[89] Magliano and Zacks (2011).
[90] Smith, T.J. (2012).

The idea that it is possible to have an exclusively propositional experience of a movie has been abandoned in favour of the theory that assumes forms of corporeal engagement through which we access the plot and "move" freely within it, without losing sight (in fact, quite the opposite!) of the world vision which this particular perspective now makes possible.

The next chapter examines the question of spectator motor involvement, analyzing various sequences from well-known movies and basing the analysis and interpretations on the embodied simulation perception model.

2
Stilted Movements and Improbable Stares

Someone moved, but who?

Alicia, daughter of a German spy, played by Ingrid Bergman, received orders from Devlin, an American secret agent played by Cary Grant, to foil a pro-Nazi plot woven by a group of fearsome characters led by Alexander Sebastian (Claude Rains), an old friend of Alicia's father. For the plan to succeed Alicia must follow Devlin to Brazil, where Sebastian lives, and insinuate herself into the criminal's life to the point of becoming his wife. Alicia and Devlin are on the verge of falling in love but are aware that a relationship could compromise the success of the mission; to the torment of this melodrama, add the tension generated by the charade of Alicia's daily life and the diffidence and suspicions of Alexander's mother, who does not trust her daughter-in-law and does not intend to renounce her castrating influence over her terrible son.

This is the story behind *Notorious* (1946), the masterpiece by Alfred Hitchcock that François Truffaut defined as "the quintessence of Hitchcock's cinema" in his book-length interview.[1] What strikes us most in films like *Notorious* is Hitchcock's almost complete indifference to the plot; he is not interested in what the pro-Nazis want, even less in the terrible projects for the uranium in the wine bottles Sebastian is hiding in his cellar. He is not even interested in examining Alicia's complex psychology; Alicia who has seen her father arrested, condemned, and commit suicide, has fallen in love with an American secret agent, and has married for a good cause (but recanted with god speed) one of the world's most wanted criminals.[2] No, what is important to Hitchcock is that the movie should exude danger in every shot, and that the audience is kept in a state of suggested anxiety, not so much by the plot itself as by the behavior of the characters and particularly by how this behavior is presented by the movie. In other words, the state of suspense in which we find

---

[1] Truffaut (1993, p. 139).
[2] See what Hitchcock himself says to Truffaut about the plot of *Notorious* and the search for the "MacGuffin," the simple and somewhat absurd pretext that sustains the narrative (Truffaut, 1993, p. 139–40). Some passages in the film's temporary screenplay, dated 1945, might have given more weight to the underlying historical and political events. See Alonge (2012, pp. 218–19).

The Empathic Screen: Cinema and Neuroscience. Vittorio Gallese and Michele Guerra, Oxford University Press (2020). © Oxford University Press.
DOI: 10.1093/oso/9780198793533.001.0001

ourselves at every viewing of *Notorious* has nothing whatsoever to do with the story; if we were to analyze the plot we would find the movie somewhat disappointing. It is the web of actions, emotions, and sensations that Hitchcock masterfully weaves with a precise schema of camera movements and editing that keeps us on the edge of our seats and feeds us with a perfectly calibrated level of stimulation that goes beyond (or rather, remains prior to) our cognitive approach to the movie. Even if we have seen *Notorious* ten times over, we are still gripped with uncontrollable anxiety when watching the two or three scenes of maximum tension; what is the explanation for this?

David Bordwell referred specifically to *Notorious* when proposing his "firewall hypothesis," according to which much of what happens on the screen and which has to do with the intensification of our involvement depends on very low-level elementary processes.[3] Close-ups, rapid camera movements, cuts, and audio effects are all techniques of sensory immersion that we respond to pre-reflexively, without the need to cognitively process what we are seeing. According to this interpretation, one of the ways that filmmakers exert their power over audiences is their use of this precognitive and low-level potential, at the basis of which lies the true secret of the magnetism of cinema. In fact, looking at the works of many theoreticians and commentators of the first decades of the 1900s, when the explosive force of cinema was increasingly impacting millions of spectators thronging movie theaters the world over, we find they laid great emphasis on the strength of the sensory shock of this medium; indeed, it was considered to be the most impactful of its characteristics.[4] Some early theorists, including Victor Freeburg, were particularly explicit regarding the primary relationship that anticipated the comprehension and interpretation of what we see on the screen. Freeburg himself, writing of the fascination that the moving images and the characters of a movie exert over spectators, observed that "This response of our senses to human form and physical movement is primary and elemental, and takes place before our brain has time to interpret the dramatic significance of the visible stimulus."[5]

Bordwell's firewall hypothesis aims at explaining why, when we watch a thriller, a horror movie, or even a particularly moving melodrama for the umpteenth time, we still experience the same emotions as we did during the first viewing, even though we know the movie like the back of our hand. One explanation for this could be that our ability to control negative emotions is

---

[3] Bordwell and Thompson (2011, pp. 99–101).
[4] See Gallese and Guerra (2012, 2014).
[5] Freeburg (1918, p. 12).

limited; another that intersubjective forms acting non-linguistically activate specific relationships each time with what we see, decoupled from what we already know.

Talking of facial expressions that make many close-ups meaningful or particular actions performed by the characters in a movie, Bordwell suggests that mirror neurons could facilitate this form of primary relationship between the spectator and the movie. What we are proposing is that the motor system plays a decisive role not only with regard to the firewall hypothesis, but also on a more general level in the embodied relationship with moving images; moreover, that the starting point for the study of the intensity of our experience at the movies is the understanding of intersubjectivity based on motor interaction. We do not mean to say that what we experience when viewing a movie can be fully understood insofar as it is attributable to forms of motor resonance that occur between the spectator and the film. Our objective is more minimalist and, possibly, less ambitious. Our hypothesis is that these forms of motor resonance play a role that is by no means secondary in the coordination of the systems of symbols and meanings constructed by cinema and that the theories linked to embodied simulation can also be usefully applied when analyzing films.

As an example of this, we will take one of the most memorable "peak" scenes of *Notorious* for which the spectator's level of involvement has to be particularly intense. Devlin has asked Alicia to steal her husband's key to the cellar where he keeps the bottles of uranium. Sebastian never lets his keys out of his sight; Alicia therefore has to take advantage of a moment when his attention is momentarily distracted to slip the key off the key ring. This is a critical point in the plot, and Hitchcock organizes the decisive sequence with impeccable style (Fig. 2.1). It opens with an establishing shot of the outside of Sebastian's villa. It is evening, the lights are on. Hitchcock draws us into the interior of the villa where we find Alicia in evening dress, tinkering with her earrings. The camera follows her until she stops just a step from the lens and we have a full-face close-up shot. Her gaze is focussed off-screen and the next shot shows the door to the bathroom, off Sebastian's bedroom. The door is slightly ajar and we can see a shadow projected on the door. The shadow is moving. We know that this is her husband's shadow, he is in the bathroom, getting ready for dinner. Hitchcock swings the camera back to Alicia, who takes a small step forward and lowers her gaze to another point off-screen. The next shot takes us back to the bedroom with the bathroom door and the moving shadow, but this time the camera moves forward and seems to bring us close to a table on which rests a set of keys. The camera movement is complex, veering slightly towards the left and as it proceeds, beams gradually down, zooming in on the table until

56 STILTED MOVEMENTS AND IMPROBABLE STARES

we can see every detail of the keys. Just when the key ring seems to be within our grasp, a cut swings us to a full-figure shot of Alicia, still standing on the threshold of the bedroom. Finally, she moves over to the table and starts to extricate the key from the key ring under our very eyes, while Sebastian's voice reaches us from off-screen, piling on the tension; he can't see her, but he speaks to her, showing us that he is aware she is in the room.

The problem that Hitchcock had to solve in this complex sequence was how to bring the spectator to an almost unbearable level of suspense, playing on the risk of Alicia being discovered by her husband—whose moving shadow on the door represents a constant threat—in the act of slipping the key off the key

Fig. 2.1 *Notorious*, Alfred Hitchcock, 1946.

Fig. 2.1 Continued

ring. The editing that connects Alicia's gaze (and intentions) to the shadow has created an anguished relationship that develops in cognitive terms: We know that Alicia is looking for the keys, we know that Sebastian is in the bathroom, we wonder if she will have enough time to do what she has to do. Classical editing, based as it is on precise connections and forms of erotetic narration as Noel Carroll would call them, makes use of inference and deduction, aiming at a perfect closure and framing of the plot.[6] However, if Hitchcock had

---

[6] Carroll (2007).

been content in this case to use a cognitive mechanism based on the classical Hollywood style (which had always guaranteed a certain economy of vision, oriented towards the mediation created between the camera and the editing process being as transparent, and as easy to comprehend, as possible), our level of involvement would not be nearly so high. Of course, we would still have been on the edge of our seat, concerned about Alicia's fate, and if the filmmaker had played on the connections between glances and cuts he might have been able to create a surprise effect, but this would not have been sufficient to create suspense.

This is why Hitchcock uses camera movement; it is this movement that creates the overpowering tension. He builds it up like a false point-of-view shot, making it as sinuous as possible so that the spectator naturally and immediately associates it with Alicia's movements. Nothing more is needed; we attribute the immanence of a human body to that movement. Hitchcock conveys this impression even more clearly when he moves the camera to simulate the gesture of the keys being picked up; in that precise moment, the spectator has the impression that the keys have been picked up and the difficult task entrusted to Alicia has been successfully completed. The camera itself took over the motor intentions of the actress in a highly charged emotional scene, in which our anticipation of what Alicia will manage to do comes into potential conflict with what we think her husband Sebastian will do. Will Alicia manage to get the key off the ring before Sebastian comes out of the bathroom? Would we be able to do it if we were in Alicia's shoes? According to our hypothesis, the possibility of making these apparently cognitive valuations lies in the predictive nature on which the functional architecture of the motor system is based; as this system is activated in simulation mode, it does not produce any actual movement but gives us the possibility of anticipating and understanding the movements made by others.

This motor simulation has drawn us so far into the quick of the sequence that when Hitchcock reveals that the movement is a mere fantasy (nothing more than the actress's—and our—mental projection) we are beset by feelings of frustration. Alicia didn't move closer to the keys; she is still there, motionless, on the threshold of the bedroom. The tension that was created by the focus on the keys is defused and is replaced by an even stronger sense of suspense. It is certainly not of secondary importance that our viewpoint, from which we see Alicia standing in the doorway, is exactly the same point to which we were drawn by the camera movement. Someone has reached the table, the camera movement has indeed produced a change in position, but it is the spectator who moved, not the actress. Once again Hitchcock emphasizes the impotence

Fig. 2.2 *Lady in the Lake*, Robert Montgomery, 1947.

of the spectator compared to the possibilities of action of the actors. To all intents and purposes we are there with Alicia ... but there is no way we can pick up those keys.

This narrative strategy—this challenge to the continuity and transparency typical of classic American movie-making through more perceptively and stylistically complicated long shots—is often to be found in the movies of the 1940s and there are many examples from a limited cluster of years that could be cited.

Two "experimental" films came out in 1947; they developed a first-person narrative style totally anchored to the subjective viewpoint of the protagonist. In his *Lady in the Lake* (Fig. 2.2), Robert Montgomery applied this style to almost the entire movie, not only making viewing difficult, but also failing in the attempt to sharpen the projection and the spectator's sense of identification with the leading character, who is seen only on the rare occasions when he moves in front of a mirror, a ruse used by Rouben Mamoulian in his 1931 rendering of *Dr Jekyll and Mr Hyde*. As Barthélemy Amengual observed, the leap is a leap too far; the image is no longer just an image, it is *seeing*, and the filmmaker is no longer able to guarantee that his spectator will have that level of "satisfaction" from watching the actor, nor can he create transfert and connections that obviously have no place in a movie that "has taken on itself to show only long-shots, doors, stairs and stains."[7]

This experiment shows that the spectator is not able to identify primarily and for an extended duration with the character's gaze; he needs to see the actor from a third-party perspective, to be clear about his spatial location in the movie, to understand his relations with the other characters, his expressions and movements. He needs these elements in order to be better able to valuate not only the character's behavior, but also his thoughts and emotions. It is well

---

[7] Amengual (1971, p. 87). See also Chateau (2011, pp. 49–50).

known that the voyeuristic desire expressed by cinema is a decisive element in our experience as spectators,[8] and although it is reinforced by point-of-view shots (even when false) and other solutions such as when the character gazes into the camera and seems to communicate directly with the spectator, these do represent a challenge to the standard of film viewing and as such must be used with care. In Amengual's opinion, Montgomery's movie "is interesting because it draws limits to the *physical* subjectivity of cinema."[9]

In the same year as Montgomery's *Lady in the Lake*, Delmer Daves used the same idea in *Dark Passage* with greater success. In this case, however, the filmmaker adopted the convincing expedient of limiting the first-person narrative style to just one part of the movie: A fugitive from San Quentin, forced to make a new life for himself to avoid capture, has a surgical operation to change his physiognomy so that he won't be recognized. The spectator participates in all the phases of the flight through the eyes of the fugitive and is "released" only when the operation is over and the jailbird looks in the mirror to find...he has the face of Humphrey Bogart.

This example of "first-person camera" or "Camera-I" as it was called at the time, was quite efficacious, limited as it was in duration. Of course, at certain points the camera movement was not sophisticated enough to mimic the twisting of a human head, in others the sense of subjectivity bordered on the ridiculous: When the character of the movie, Vincent, finally found someone to give him a lift, the whole scene is concentrated on the profile of the driver, when, of course, no passenger would ever sit and stare fixedly at the driver of a car, and certainly not if the passenger is a fugitive from justice. In spite of this, the first-person camera technique was successful here and when we ask students if they see anything strange in this sequence, hardly anyone remarks on Vincent's improbable stare. In October 1946, Delmer Daves filed a written document with Warner Bros, entitled "Observations on the camera acting as a person: *Dark Passage*,"[10] explaining that he was experimenting with the idea of transforming the camera into a body so that the shots projected onto the screen would represent the character's visual field. Every time the camera moves it means that the character is moving, turning his head, crouching down to hide behind a bush or along the wayside, walking warily, or speeding up. He recorded concern with the vibrations that are difficult to control in a hand-held Arriflex that has to be kept at eye level to ensure that the shots appear to be

---

[8] Brown (2012). See also Dagrada (2015) on the point-of-view shot.
[9] Amengual (1971, p. 87).
[10] Sobchack (2011, pp. 72–6).

 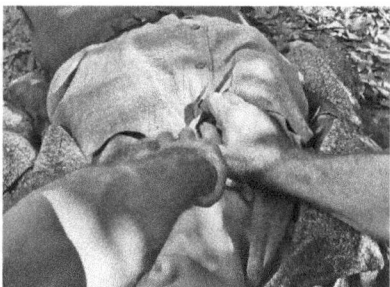

Fig. 2.3 *Dark Passage*, Delmer Daves, 1947.

from the character's viewpoint, about the downtimes which make the fixed stare of the camera so stilted (the example of Vincent staring at the driver who gave him a lift is a case in point).

Quite often Vincent's hands appear in his visual field (in the shot, that is), sometimes even his forearms and feet in order to give the idea of a complete body, of which during the movie, as in real life, we see almost everything except the face (Fig. 2.3). Exploiting the movements of the camera, Daves translates the stability of the character into motor sequences as if he were animating Ernst Mach's first-person perspective drawing (Fig. 2.4).

When studying the implications this style has for the representation of subjectivity in cinema, and also the comparison of the coefficient of subjectivity attributed to the camera by the other actors in the movie and the spectator himself, Vivian Sobchack notes how oppressive this form can be, as the filmmaker has reduced the body to nothing more than its immanence, and condemned it to remain outside the shot in the play of real and potential glances aimed in its direction, attempting to give it the dignity of being the subject.[11]

Simulating the movements of the character is a complicated business: At the very least a Steadicam would have been necessary to make the relationship with the spectator more fluid and involving. The intensification of the intersubjective relationship that is so clear in the sequence in *Notorious* is weakened here both by the duration (a spectator cannot be expected to maintain such a lengthy and radical identification with a fictional character) and by the technical difficulties inherent in reproducing the gestures and intentions of the character with the movements of the camera. We are not talking here of a single movement, bursting stylistically and narratively into a consolidated situation;

[11] Sobchack (2011, p. 74).

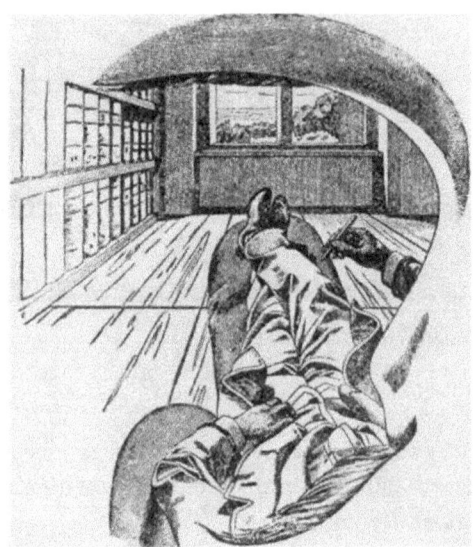

Fig. 2.4 Ernst Mach, *Subjective Perspective*, 1886.

we are talking of continuous stimulation that at a certain point exceeds our possibilities of holding on solidly to such a strong form of identification.

As Francesco Casetti said, the filmic experience is a combination of states of *excess* and *recognition* practices; the first are elements that impact us directly, and exceed our ability to control them rationally, while the second not only provide us with the exact notion of what we are experiencing, but also with the ability to express and manage the various meanings of the experience.[12] The dialectic between excess and recognition greatly contributes to modulating our identification with the movie. In fact, it appears that a clear segregation between the filmic space and the space occupied by the spectator is a sine qua non condition for identification; while promoting the impression of reality, this segregation simultaneously presupposes a clear division of space. As we will see later in this chapter, this issue is expertly dealt with by Buster Keaton in his *Sherlock Jr.* (1924): When the character/spectator appears as if by magic on the screen, becoming part of the narrative in spite of himself, his troubles are just beginning owing to his total inability to anticipate the actions that unravel as the movie progresses. We cannot anticipate what is contemplated on the screen, and this throws us. It is therefore of the utmost importance for a

---

[12] Casetti (2009, p. 57).

filmmaker to be able to adjust this dialectic, to understand where the elements of excess come into play and where it is necessary to trigger the recognition of the fictional element in order that the spectator's identification is not diminished.

This is why sequences in which it is possible to establish heightened forms of corporeal involvement lose their effect if they are not strengthened and framed by sequences that put the spectator in a position to clearly recognize his externality to what he is observing and the modalities used to achieve it. It is also why our approach cannot be readily applied to the complete filmic experience but can be very useful as an instrument for analysis in peak sequences, those sequences used in almost every movie, frequently associated with the "recognition" system adopted by filmmakers and screenwriters and which remain longest in our memory.

Although neither *Lady in the Lake* nor *Dark Passage* succeeded in stabilizing this type of solution for periods of time that exceeded the temporal limits of the "excess," there were other films in those years that did, using more limited forms of motor identification linked to false point-of-view shots. *The Spiral Staircase*, one of Robert Siodmak's best known movies, which debuted in the same year as *Notorious*, is one such film.

The plot is built around a psychopath who murders young women with physical disabilities in the New England of the early 1900s. Helen (Dorothy McGuire), a mute, is working as a governess in the residence of the Warren family, composed of the elderly bedridden mother (Ethel Barrymore) and her two sons, Albert (George Brent), a respected professor, and Steven (Gordon Oliver), a philanderer. Because of her disability, Helen is a potential victim of the serial killer and Mrs Warren is concerned for her safety as the murderer lives in that very house…

As can be imagined from the brief summary above, this is a nerve-wracking thriller, constructed on a number of archetypes such as the "big old house" and which evolves, as many of the best thrillers do, in a dark, stormy night.

Siodmak started his career as filmmaker in Germany, shooting films that from both the visual and the psychological point of view are on a par with the great works of Fritz Lang or Georg Wilhelm Pabst, and which are imbued with the last breaths of German expressionism. In fact, his American film noirs stand out for their fortunate conjugation of German models with a genre that from a purely formal point of view was receptive to the German tradition. *The Spiral Staircase* is a perfect example of this; while the German model is evident in many instances of cropping, in the lighting techniques used to create a terrifying atmosphere and certain narrative concepts such as the detail in the murderer's eye, the strategies used to create suspense are completely in line with what was the fashion of the time in Hollywood.

Take, for example, the scene in which Siodmak lets us glimpse the danger threatening Helen and reveals the menace hidden in the Warren residence. This is one of the movie's peak sequences (Fig. 2.5). Helen climbs the great stairway that connects the living area of the house with the bedrooms above; she stops on the halfway landing, in front of a huge mirror. Every care has been taken with the scenography; two lamps flicker beside the mirror, in which we see reflected the dining room below with the great table on which two lit candelabra have been placed. Helen looks at her reflection in the mirror, tidies her hair, and then tries to speak but fails (we discover later in the movie that her mutism is due to a childhood trauma); disconsolate, she lifts both hands to her throat in a gesture that should be interpreted as a desperate attempt to understand why she can no longer emit sounds but to the shrewd spectator also suggests the risk of violent death, probably by strangling.

It is at this very point that Siodmak decides to cut, and the next shot, taken from the top of the stairs, shows Helen in full figure. This type of unjustified cut is interpreted by the spectator as the subjectivity of someone spying on the character and, given the genre and the atmosphere, it is rational to suppose that that someone is her potential murderer. How can this sensation be emphasized to the point of excess? Just like Hitchcock, Siodmak decided to play on the spectator's motor identification. The camera starts to move to the right, not only simulating the movement of the observer from one point to another, but also by lowering the view point, giving the impression of crouching behind the banisters to keep out of Helen's line of vision. The movement continues for quite some time, keeping Helen in the shot, visible through the banisters; the soundtrack heightens the atmosphere of danger, emphasized by a clap of thunder, but just as the movement slows down and it seems that the killer has found the ideal hiding place in which to lie in wait for his victim, two legs suddenly appear, filmed from the waist down. The identification effect and the coefficient of subjectivity of the camera movement are disintegrated by this apparition: No-one was spying on Helen from that viewpoint and the killer was already hiding in the corner to which the camera had brought us with all that tension.

This scene has a number of elements in common with the better-known sequence of movement in *Notorious* that we analyzed earlier; above all, the false point-of-view shot is reinforced by the camera movement, which is also to be considered false if we think of it as consubstantial to a moving subject, but it is, of course, an actual movement. It is intended to imitate the movement of a

Fig. 2.5 *The Spiral Staircase*, Robert Siodmak, 1946.

human being, in this case the movement of the potential killer, in hiding while spying on the central character of the movie. Just as happened in *Notorious*, someone did indeed move, and, while they were moving, continued to keep Helen in their sight. Hitchcock inflicted a sense of frustration on the spectator who had been lulled into believing that he had managed to grasp the keys; Siodmak's spectator lives a moment of anguish as he "sees" the scene through the eyes of the killer, crouched in a menacing position by the banisters, only to be filled with terror as he realizes how close he is to the real killer. This surprise twist is just as successful as Hitchcock's when the camera shows Alicia motionless at the bedroom door.

As Hitchcock has taught us, when shooting a thriller, the filmmaker can choose whether to use suspense or surprise. If he opts for suspense, he will have more time in which to play on his spectator's fears and sense of anguish; if he opts for surprise, he will spring something unexpected on his unsuspecting spectator, of undoubted effect but of much shorter duration. To create a perfect situation of suspense, the spectator must be well informed of what is happening as is the case in both *Notorious* and *The Spiral Staircase*; the spectator was fully aware of the task entrusted to Alicia and the risks that Helen was running. All the same, at a certain point at the height of the two key sequences, suspense is rekindled by the surprise (the spectator did not expect that Alicia would still be standing in the doorway, or that the real killer was hiding in a different point of the landing) generated by the resonance produced by the camera movement. Of course, the lighting, the shadows, the choice of ambience, the soundtrack, and the acoustics all contribute to the mood of the scene, but it is the camera movement that captures the spectator's imagination.

The sequences we have analyzed here and their ability to transmit the motor intentions of the characters and guarantee a response in the spectator lead us to talk about alignment, a technique used by filmmakers to create identification between the character and the spectator.

Talking from the cognitivist point of view, Murray Smith observed how alignment develops through two interlocking functions—spatio-temporal attachment and subjective access—which require models of physical and psychological coupling with the character.[13] The objective of the structures of the alignment is not only to facilitate the regulation of the narrative competences, their circulation between actors and spectators, but also the agentive relationship between the actor on the screen and the spectator sitting in front of the screen. Talking about perceptive alignment, Smith notes that the point-of-view shot is only one of many possible narrative techniques the filmmaker can use to convey the character's subjectivity and, notwithstanding the fact that it allows us to share the position and point of view, the spectator is often disoriented by the limits to the fundamental narrative structures on which the movie is constructed, among which is the identification with characters that are actually seen in action.

In his cognitive analysis of subjectivity Smith talks of what he calls the "fallacy of POV [point of view],"[14] which drastically reduces the information available to the spectator and forces him to infer meanings from superficial details

---

[13] Smith (1995, p. 83).
[14] Smith (1995, p. 156).

that fall within the point of view and do not possess the systematic clarity that is typical of objective narration.

More often than not, the camera plays out its mediation role defining a context of action that is fairly vast and regulated by top-down interpretation mechanisms that rotate around three fundamental variables: The character/agent, time, and place where the action is being performed.[15] Even subjectivity is always structured in this way: It requires a person who perceives a part of the actual situation (and the spectator has to be informed of this in some way) and that person (who will then be a "victim" of the omission presupposed by the subjectivity) must be inserted into that situation in that particular type of image that, as Amengual would have said, becomes a glance. Now, in extreme cases, such as in *Lady in the Lake*, the fallacy of the point of view introduces the failure, or at least the weakness, of the mediation and the representation of an agentive subjectivity. However, in *Notorious* and *The Spiral Staircase*, the point of view (a false point of view in both cases) worked perfectly because it was framed by the various alignments that regulated the narrative mediation and made it comprehensible. As well as providing information on the situations in which Alicia and Helen found themselves, the filmmaker allowed the spectator to empathize with them through an appropriate use of close-ups and setting the camera movement in the core of a sequence shot adhering to the rules of transparency and continuity of the classical style. This lasts for the entire sequence and is emphasized by the camera movement. In that moment, a shot is created which

> ... while it does not reproduce precisely what the characters see, it still suggests something of their manner of experiencing the world. This something is the sensation of direct contact, a trace of curiosity, a mental attitude of availability and interest—in sum, those aspects that in relation to the world are complementary to the gaze.[16]

The fact that the shot is animated by a movement that goes beyond the considerations of Smith and Branigan and throws a different light even on what Casetti said—our frustration when something doesn't happen increases in proportion to our level of identification—imbues this availability, this intentionality with significance. In fact, our identification with the scene of a movie, and particularly in thrillers, is greatly influenced by the degree of simulation that

---

[15] Branigan (1992, p. 159).
[16] Casetti (1998, p. 70).

the images are able to evoke in us. In this regard, we maintain that the functional mechanism of embodied simulation expressed by the activation of the diverse forms of resonance or neural mirroring discovered in the human brain play an important role in our experience as spectators. Our ability to share attitudes, sensations, and emotions with the actors, and also with the mechanical movements of a camera simulating a human presence, stems from embodied bases that can contribute to clarifying the corporeal representation of the filmic experience supported in the phenomenological field[17] over recent decades, without seeking forms of dialogue with neuroscientific evidence or models.

Criticizing certain theories of subjectivity, George Wilson (agreeing with observations by Kendall Walton) sustained that subjective shots do not give the spectator the possibility of imagining to be at one with the character nor of occupying the same vantage point in fictional space; according to Wilson they simply allow him to consider the vantage point as the position the character pretends to take in the movie and from which the camera shows the spectator his viewpoint.[18]

Now, while Wilson's interpretation may appear acceptable from the cognitive point of view, as soon as we start evaluating other forms of subjectivity codification the situation changes; for example, the "privilege" of the vantage point may be achieved by the specific choice of movement that evokes forms of greater resonance and complexity in the spectator.

Notwithstanding the total externality, the spatial segregation of the symbolic content vis-à-vis the spectator—in this specific case the hiatus that separates the spectator from the screen—the connection with the body is maintained on a multiplicity of levels. This connection is established not only with the body of the actor on the screen, but also with the body of the filmmaker (from his presence on the set, to the way in which he directs his/our gaze in the narrative space of the film). At the same time, we have seen that film/spectator separation is necessary for there to be full involvement with the planned narrative/perceptive material, even if the spectator does remain firmly anchored mentally and physically to the prosaic everyday world (the seat from which he watches the movie after paying his ticket). Now, if that seat were to give way under his weight, that other space—the space belonging to the narrative fiction in which he was totally absorbed before the unfortunate incident happened—would

---

[17] See the classical studies by Sobchack (1991, 2005). Barker (2009) applies the same methodology, while Shaviro (1993) and Bellour (2009) start from different premises. MacDougall (2006) also worked on body-image intertwining.

[18] Wilson (2006, p. 84).

immediately disintegrate and he would be left with the prosaic task of getting to his feet.

The transition between these two states, so distinct and easily distinguishable from the point of view of our experience, is by no means equivalent to transiting from one alien world to another. The body quivering in trepidation beside Alicia and Helen is ours, as is ours the body that contemporaneously permits us to live this experience from our seat in the cinema. This same body gives us the possibility of participating contemporaneously in more than one reality.

Resonating with movement

In our view, this is the precise point where the contribution of cognitive neuroscience can deploy its heuristic power, giving us the tools to develop new analyses of film style and mechanisms that are efficacious in "capturing" the spectator. The sequence in *Notorious* that we examined earlier will come in useful for evaluating these aspects. The particularly complex camera movement used to create the hypothetical path that Alicia should take to reach the object of her desire (the keys to the cellar), renders those keys increasingly desirable as the distance diminishes, so guaranteeing our heightening involvement as they enter our peri-personal space, that space in which they are within our reach. In the next chapter we will show that this heightening of our involvement occurs due to the simulation of the movement that we would make to reach the keys.

That "within-reach" space, the *"zu-Handen"* space of which Martin Heidegger wrote, is mapped in our brain by the same parietal–premotor motor centers that guide the actions of orientation or grasping in that space. Premotor neurons instantiate an embodied simulation mechanism that maps what touches us, what we see or hear in our immediate vicinity, on the motor potentialities of our body. The table is also the paces needed to reach it, just as the keys of the cellar are the potential act with which we grasp them. Like Alicia, we map the position of the table on which we can see the keys on the basis of the motor potentialities expressed by our brain–body system, which here we have analyzed exclusively from the point of view of the actions involved, which, of course, does not mean to diminish the crucial importance of emotions and desires in planning our actions.

Now, to get back to Alicia (and to us). We are moving toward the table on which the keys have been left. As soon as those keys enter our peri-personal space they become the potential object of a grasping-by-hand action. Alicia, and we with her, can finally grasp those keys. What happens in this precise moment in our brain? Among other things, the parietal–premotor circuits that contain canonical motor neurons are activated. The canonical neurons in the macaque and their counterparts in the human brain "translate" the three-dimensional shape of a graspable object, such as a set of keys, into the motor program we need in order to perform the action of grasping.

The whole point is that the activation takes place even if we just look at the objects within our grasp without any intention of literally performing the action. We are simply simulating the action of grasping the keys, our muscles are not activated and we do not make any movement. In our hypothesis this motor simulation contributes significantly to our degree of identification with what is happening on the screen. This approach gives us the possibility of saying something more than just a reference to an undefined mechanism of empathic involvement with the action narrated in the movie, based on an equally undefined embodiment mechanism with the scene.

This approach can hypothesize on which neurophysiological mechanisms this particular "immersive" aspect of the spatiality and intentionality of the actions in the movie are based; actions that can involve us, bore us, or disgust us, and with which we can identify (or not, as the case may be). The advantages are that not only can this approach be empirically verified, but it can be applied, positively and negatively, to a variety of filmic situations that go far beyond what the Italian poet Ugo Foscolo termed "the correspondence of loving senses" *("le corrispondenze d'amorosi sensi")* with the characters of the movie.

Positions

The ability of cinema to produce meanings and symbols that the spectator is able to process without difficulty is based almost exclusively on stylistic solutions of the type we have just seen. Apart from the cognitive processing involved, these meanings and symbols develop from non-linguistic forms of relationships based on the embodied reception of film, entrusted to the behavior and movements of the camera and the editing. What Mark Johnson wrote regarding the "bodily depths" of our corporeal encounter with the environment can certainly be applied to our relation with the images and fiction that we encounter with increasing frequency in our everyday life.[19]

---

[19] Johnson (2007). Recently Johnson wrote the introduction to the book edited by Coëgnarts and Kravanja (2015).

The issue is to understand to what extent the expressive ability of a work of art, its power to involve the observer/spectator, to communicate (in the case of a movie, this communication can last for extended periods of time, even a number of hours), is modulated by its embodied reception. We want to understand how it evolves by taking into consideration the body, gestures, and models of physical and carnal contact to be found in the artwork itself and which are to a degree presupposed by the techniques used. This theme is by no means recent. The dialogue that Sergej Ejzenštejn entertains with the theories of Meyerchol'd on biomechanics,[20] developed on the basis of a profound discussion regarding the body as a generator of meaning, and movement as the core of a multitude of meanings, with the power of having an immediate impact on the senses of the spectator. As Antonio Somaini noted on this score, at the base of Ejzenštejn's thoughts on these themes lies the necessity of directing them to "the *expressivity of artwork in general*, which is just as applicable to theater as it is to cinema."[21] It is certainly no coincidence that the focus has moved from considerations regarding the physical body of the characters to the motor engagement required from the spectator and modeled by the movie.[22]

How does the development of new technologies produce forms of *embodiment* or *re-embodiment*? Don Ihde suggests that the media should be studied starting from the technical characteristics that connect us with situations that can be described as being "in our reach." Ihde refers specifically to cinema in order to explain how unisensorial (visual) simulation evokes sensations that he defines as apperceptive, capable of amplifying our experience. He takes as an example the legend of the reaction that the train in *L'arrivée d'un train en gare à La Ciotat* (Lumière Brothers, 1895) evoked in spectators. The movement of the train, which Ihde describes as "virtual," produced visible motor reactions in a number of spectators, showing the primacy of movement and its forms of resonance in embodied relations. Ihde observes that many contemporaneous technologies give us "special powers" and incorporated possibilities that we otherwise would not have, but at a price. It doesn't matter how deeply we embed these technologies, we are destined never to forget, nor to abandon the real position from which we exploit/enjoy these worlds that are offered to us.[23]

---

[20] Ejzenštejn (2009, pp. 89–90).
[21] Somaini (2011, p. 9—our translation).
[22] Essential reading on these subjects is the essay Ejzenštejn wrote in October 1924 entitled "Montage of attractions," in which he analyzes the phases of motor contagion induced by the reproduction of the movements of the agent by the observer and in which he discusses the modellng of the "motor variant" in the dynamic schema of the film. (Ejzenštejn, 1986, pp. 227–50).
[23] Ihde (1979, pp. 142–3). See also Ihde (2002, 2010).

Cinema has redesigned our way of interacting with these worlds and without a doubt has left a mark on our understanding of the virtual intersubjectivity typical of many of the so-called new media. In 1927 Walter Benjamin wrote that a *"new realm of consciousness"*[24] came into being with film; he did not just mean that cinema would have social and political effects that would reach far beyond imagination, he meant that it would launch the spectator into a new experiential dimension (a new realm, a new *ontology*) on multiple levels that would require a reconfiguration of the spectatorial regimes and practices conceived up till then. He saw film as a prism ("the only prism") through which modern man finds his immediate environment, the spaces in which he lives, laid open to him. "To put it in a nutshell, film is the prism in which the spaces of the immediate environment—the spaces in which people live, pursue their avocations, and enjoy their leisure—are laid open before their eyes in a comprehensible, meaningful, and passionate way." The spatial–temporal transformation that film performs gives a new form to our daily environment; it also fine tunes our ability to move in fictional worlds, increasing, together with our impression of reality, our psychophysical involvement with the virtual space that is typical of the big screen.

Benjamin just glimpsed this, but glimpsed it with extreme clarity (and brought it into even greater focus in his later work, *The Work of Art in the Age of Mechanical Reproduction*); however, it was already nascent in earlier reflections on film and frequently returned like a Gordian knot in many diverse theoretical debates.

From its very origins the cinematographic experience was remarkable for the strength of the sensorial impact on its spectators. Fear and wonder are the most frequent sentiments, and they are capable of provoking peaks of agitation that the spectator has difficulty in controlling. In 1907 the Italian author Edmondo De Amicis wrote a short story significantly entitled *Cinematografo cerebrale* (*Cerebral Cinematograph*) in which he spoke of an "orgy of the spirit."[25] Maksim Gorkij was even more specific; in his considerations penned in 1896 after sitting through the very first projections in the history of cinema, he noted "your nerves are strung, your imagination leads you into a strange form of life, artificially monotonous, devoid of colour and sound, but brimming in movement."[26] Cinema produces an intensification of the spectator's nerves and leads him into a strange dimension (and here we have again Benjamin's new realm

---

[24] Benjamin (2005, p. 17).
[25] De Amicis (1907, p.46).
[26] This text is part of an anthology in Banda and Moure (2008, pp. 52–7; our translation).

of experience), in which a mechanical device has made everything that is ontologically fragmented appear uniform; and while there are no colours or sounds in this new space created by cinema, there is an overwhelming impression of reality, which, according to Gorkij, is due to the fact that the images are "brimming in movement." Gorkij even goes so far as to say that cinema "is life itself, it is movement taken on the quick."[27]

The most fascinating element of all is still the question of forms of involvement. Various theoretical writings of the early decades of the 1900s insist on the immersive properties of cinema and later, when comparative studies were the fashion—in the years in which the specificity of the new art form was sought through comparison with other forms of expression—many saw sensuality as the distinguishing feature. Take, for example, Antonin Artaud, who, in 1923, wrote that there could never be more than just one genre of film, "the kind in which all the means of sensual action available to cinema have been utilized" because cinema is "a remarkable stimulant" that "acts directly on the grey matter of the brain."[28]

Some years later, Erich Feldmann, an exponent of French filmology,[29] described the great mystery of the power that cinema exerts:

> The film requires the spectator to do something that at first sight seems impossible: to transport himself into an unreal situation without the aid of stimulants, drugs, or psychophysiological modifications produced by the luminous projection alone, all the while remaining vigilant in the reality of the cinema hall, believing in the reality of the film that absorbs him.[30]

The spectator dwells in the real world (Feldmann, in fact, uses the French term *demeurer*), but at the same time he is able to let himself be transported into another dimension and find his bearings there, to move confidently in this fictional space, completely absorbed in the virtual space-time in which he is fully immerged without having had to take any hallucinogenic substance. His self has not been altered; if anything, it may have been duplicated. This at least is what Jacinto Lageira thought, writing 50 years after Feldmann. In a collection of various contributions on the forms of subjectivity generated by cinema,

---

[27] Christian Metz agrees with this and sees in movement a clue to the impression of supplementary reality, the corporeality of objects, and above all, more directly, the perception that movement is always (or almost always) perceived as real. Movement is a "filmic manner of presence" that makes the film "extremely credible" (Metz, 1968, pp. 32-8).
[28] Artaud (1976, pp. 11-12) (in *Selected Writings*, 1976, Susan Sonntag ed.).
[29] See Casetti (1999, pp. 93-103) on filmology.
[30] Feldmann (1956).

Lageira maintains the hypothesis of duplication in almost the same words as Feldmann used:

> The spectator is the one whose consciousness may both realize the concrete and material reality of the cinematic apparatus (realizing consciousness) and project itself imaginatively into the ongoing film (imaginary consciousness) without leaving its psychophysical base, so to speak. My capacity to divide and duplicate my imaginary self explains how the base can maintain its state of sameness so that my subjectivity can simultaneously locate itself aesthetically in the film while remaining itself.[31]

In this case Lageira, citing Husserl and Sartre and taking up the Winnicottian concept of "intermediate area" or "area of illusion," hypothesizes the existence of an imaginary subjectivity that comes into play without entering into conflict with reflexive consciousness and that permits us to behave appropriately, respecting the rules of the virtual world. Scholars writing before Lageira attempted to theorize the activation of surrogate intentionality, the objective of which would be to interact with fictional and illusory worlds. Christiane Voss, for example, writes of the spectator as a "filmic surrogate body" that starting from mental and sensorial–affective resonance would "lend" the film a third dimension, incorporating the second dimension into the third[32] with the help of precise narrative and stylistic strategies. It is in that virtual surrogate space that the two dimensions are welded together, the unreal and the real as Feldmann would say, or the reflexive and imaginary consciousness of Lageira. While collaborating on the *Revue Internationale de Filmologie*, Edgar Morin wrote an important book entitled *The Cinema, or the Imaginary Man*, in which he spoke of "surrogate consciousness" regarding the flow of cinematographic images:

> This flux of images, feelings, and emotions constitutes a current of ersatz consciousness that adapts itself and adapts to it the coenesthetic, affective, and mental dynamism of the spectator. It is as if the film develops a new subjectivity that carries that the spectator along with it, [...]. *Cinema is exactly this symbiosis: a system that tends to integrate the spectator into the flow of the film. A system that tends to integrate the flow of the film into the psychic flow of the spectator.*[33]

---

[31] Lageira (2011, pp. 150–1).
[32] Voss (2011).
[33] Morin (1956).

The crux of the matter continues to be the relationship between what happens on the screen and what really happens to the spectator: Morin goes so far as to say that cinema is like a current of consciousness that adapts and attracts the spectator's coenesthetic dynamism, as well as his mental and affective energy to the point of creating a new subjectivity that sweeps the spectator's original subjectivity along with it. He uses the term symbiosis (the flux of the film and the psychic flow of the spectator converge), offering the concept of a cinema that speaks the language of the mind (just as Hugo Münsterberg said in 1916),[34] while the mind, or consciousness, have a cinematographic element. Morin's words help us understand why certain neuroscientists who study consciousness—Antonio Damasio, Christof Koch, and Giulio Tononi, to name just a few—often turn to film as a metaphor to explain how it functions.[35] Colin McGinn would say that this is the problem of problems, the fascination that films create and use to draw us into their interior; he calls this the "mind-movie problem:"

> The "mind–body" problem is the problem of explaining how conscious experience relates to the physical materials of the body and the brain; the "mind–movie" problem on the other hand is the problem of how to explain that fact that a two-dimensional image that moves, as we experience it in a typical film, can capture our consciousness in that way and with that force.[36]

Going into greater depth, the "mind–movie problem" concerns, on the one hand, what we can define as cinema's presence modalities, and, on the other, our ability to dwell in that fictional space–time dimension through a deployment of energy generated by the type of stimulation (the images brimming with movement of which Gorkij spoke) and forms of simulation fine tuned in a brief whirl of years by what is known as "film style." This energy is not only concerned with the highest levels of our cognition; it originally springs from the lowest levels of our brain–body system. Damasio is right when he says that whoever perfected cinematographic techniques and style seemed to have been thinking of the functioning of the human brain.[37] We need to form a multimodal concept of the spectator if we are to understand how the filmic experience evolves from the forms of embodiment generated by cinematographic

---

[34] Münsterberg (1916).
[35] See Guerra (2015b).
[36] McGinn (2005, p. 5).
[37] Damasio (2008, p. 97).

techniques and makes way for the strange intersection that many philosophers and theoreticians have discussed.

Vivian Sobchack dedicated much time and energy to this point, finally coining a neologism that contains the various aspects that characterize the spectator of films, whom she calls a "cinesthetic subject."[38] She states that the adjective is constructed from three words: Cinema, synesthesia, and coenesthesia. It encompasses cinema, as a reproduction of movement, synesthesia as cross-modal perception (when the stimulation of one sense causes a perception in another), and coenesthesia as self-awareness originating from parts of the body (as we saw earlier, Morin was one of the first to express interest in the coenesthetic dynamism felt by the spectator while watching a movie).[39] The pronunciation of cinesthetic is exactly the same as that of synesthetic and so the person watching a movie is above all an individual who approaches the cinema's virtual reality from cross-modal perspectives, who is always considered in his real condition, whose self-awareness is always intact. Cinesthetic brings to mind kinesthestic, having to do with the muscular sensibility and proprioception regulating motor activity. So, we have a spectator who first of all reverberates with a form of reproduced movement ("written movement," given the etymology of the term "cinematograph") according to certain semantic and syntactical rules; while in certain cases these rules infringe the ways in which we perceive the real world or our expectations, they give the movie that "habitus of reality," as Roman Ingarden called it, which is the secret of its strong impact on the spectator.[40] In spite of the fact that the nature of cinema is basically audio-visual, and therefore stimulates our senses of sight and sound, Sobchack insists on the multimodality of the cinematographic experience, planting it deeply in the spectator's sense of real self-awareness (which is not a surrogate of either body or consciousness) and in the refusal of the separation of the senses. This position is strongly inspired by the Merleau-Ponty "rule of synaesthetics,"[41] which is one of the points of difference between the French philosopher and Husserl, confirming Sobchack's choice of position and that of much of film phenomenology.[42] And, lastly, it would be a mistake

---

[38] Sobchack (2005, p. 67 and following).
[39] The subject was also touched upon by Jean Epstein in *The Devil's Cinema* (1947), now in Epstein (2002, p. 156). See Walla-Romana on Epstein and corporeality (2012).
[40] See Angelucci on Ingarden (2012).
[41] Rodrigo (2013, p. 32).
[42] Two important works on the phenomenology of cinema were published in the US in the early 1990s by Casebier (1991) and Sobchack (1991). They represent two very different approaches to the phenomenological study of film; the first is strictly Husserlian, while the second is strongly inspired by Merleau-Ponty's carnal model. Over the course of time, Sobchack's work has predominated in film studies.

to skim over the cinesthetic element that enthralled many of the important figures of French filmology, particularly Albert Michotte and Henri Wallon, and to which Siegfried Kracauer dedicated much thought.[43]

We maintain that the theories connected to embodied simulation can offer a contribution to the formulation of a more precise definition of the potential of the "cinesthetic subject" and can constitute a base for constructive discussion regarding the dual position of the spectator that has been constantly debated by scholars and theorists of different epochs and disciplines. It must be remembered that the importance of the spectator's motor potentialities emerged both in the early days of the physiology of cinema and in certain lesser known cinematographic writings of Maurice Merleau-Ponty.[44] In 1920 two French physicians, Edouard Toulouse and Raoul Mourgue, presented the results of an experiment regarding the respiratory reactions of spectators while viewing a movie (Fig. 2.6) to the Congrès de l'Association Française pour l'Avancement des Science in Strasbourg.[45] The participants in the study were shown examples of various film genres, including documentaries, comedies, and drama, and the results lead the researchers to expound a theory that is of interest in the light of the new frontiers of film study. According to Toulouse and Mourgue:

> Film has a very strong influence on feelings and emotions, as the sense of reality generated by movement in the three dimensions of space, frequently seen in real environments, is extremely intense. What is the essential source of this sense of reality? In our opinion it is related to the motor suggestibility of the participants in our study.[46]

also

> As we know, the perception of movement gives rise to a hint of a corresponding movement [...] We believe that this is where we should start looking for one of the psychological reasons for the success of cinematographic representations with the populace.[47]

---

[43] From the days of silent movies, Kracauer was always interested in the materiality of film, the theme of movement, and their relation to human "physiological substance." See Hansen (2011, pp. 3–72, 253–79) for an analysis. Kracauer writes that "film images affect primarily the spectator's senses, engaging him physiologically before he is in a position to respond intellectually" and instigates a "resonance effect, provoking in the spectator such kinesthetic responses as muscular reflexes, motor impulses, or the like" (Kracauer, 1960, p. 253).
[44] Merleau-Ponty (2011).
[45] See Morel (2010) for a discussion on Toulouse and his work on cinema.
[46] Toulouse and Mourgue (1920, our translation).
[47] Toulouse and Mourgue (1920, pp. 143–4).

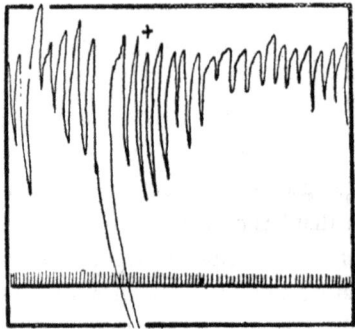 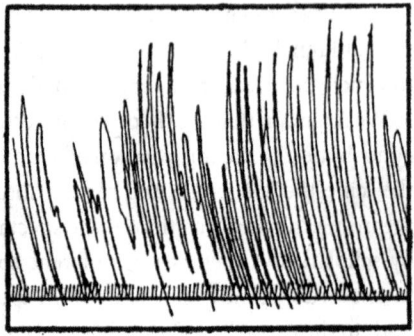

Fig. 2.6 Two graphs from Toulouse and Mourgue's experiment for investigating the emotive impact and level of attention during the projections of *Pathé Revue* (documentary, left) and the telephone scene from Abel Gance's *Mater Dolorosa* (film, right). The cross in the left-hand graph indicates a change of shot.

Reproduced from E. Toulouse, R. Mourgue, Des réactions réspiratoires au cours des projections cinématographiques (1920), "1895. Revue de l'association française de recherché sur l'histoire du cinema", 60, 2010, pp. 138–144.

The spectator's motor cognition seems to be decisive for making contact with what we could call the movie's motor schema and above all for activating an online simulation of what is happening on the screen. This latter step, to which neuroscience can, in time, provide a decisive contribution, is open to discussion and hotly debated. At present the debate is between the positions of those who consider that simulation is a metaphor of the imagination and as such can only produce offline projections (this is the position held by Greg Currie)[48] and those who think that this idea of offline simulation has to be reviewed in the terms of a theoretical proposal that takes into consideration neuroscientific research on interaction and intersubjectivity (Torben Grodal is one of the scholars who sustains this position).[49]

In the notes that Merleau-Ponty wrote for his seminar in 1953 at the Collège de France—notes that are a treasure trove of indications regarding his positions vis-à-vis cinema and also for evaluating the relationship between his line of thought and certain fringes of the filmology of the time—he makes this consideration: "therefore movement, its sense, its characteristic aspect, is perceived through the motor possibilities [*possibilités motrices*]

---

[48] Currie (1995, pp. 144–52).
[49] Grodal (2009, pp. 238–43).

of the body," possibilities that would be to movement as whiffling is to a wind instrument.[50] These motor possibilities echo (probably quite unwittingly) the *suggestibilité motrices* of Toulouse and Mourgue in the search for an interpretation of movement and its meaning, and the mutual conviction that cinematographic cognition springs from the sharing of precise motor schemas.

Going back to Merleau-Ponty's reflections on cinema, we have borrowed the phrase "grasp on the world" from his *Phenomenology of Perception*; in film the spectator comes to recognize the structures of his perception in the form of a "grasp on the world," as a "dynamic reconfiguration of his fields of experience."[51] The objective of neuroscientific research is exactly to problematize and enrich the call from objects and art forms to our pragmatic inherence, perceiving our openness to the world of art and make-believe as the expression of a potential pragmatic relationship.[52] Not only did Merleau-Ponty develop a theory of a recipient/receptive body, able to lay witness to the phenomenal truth of the movement produced by intermittent cinematographic images, but he even saw in cinema an art that would be able to contribute to overcoming the dualisms typical of western thought,[53] and proposed a converging interpretation of the world of perception and the world of expression, which when all is said and done brings us back to the mysterious duplication of the spectator. Sobchack too, used similar concepts when she said that

> [...] the cinesthetic subject both touches and is touched by the screen—able to commute seeing to touching and back again without a thought and, through sensual and cross-modal activity, able to experience the movie as both here and there rather than clearly locating the site of cinematic experience as onscreen or offscreen.[54]

In conclusion, the issues that have been raised over many decades of the theory of cinematographic spectatorship show how the theme of the positioning of the spectator can be benefited by studying the corporeality of our mediated experiences with a neuroscientific model and approach.

---

[50] Merleau-Ponty (2011, p. 119).
[51] Dalmasso (2013, p. 116).
[52] Gallese and Guerra (2013b, p. 9).
[53] Carbone (2011, pp. 112–13).
[54] Sobchack (2005, p. 71).

Sherlock Jr.

In this chapter we have discussed the mysterious ability of the spectator to orient himself and move in the space–time dimension of film, while remaining immobile in his role of observer. We have seen how by paying attention to our embodied relationship with the moving images on the screen it is possible to at least partially understand this ability and how cinema facilitates this; in spite of those stilted movements and unlikely stares, cinema knows how to draw us *physically* into the heart of the plots it wants to impart to us. The spectator's position, blocked at the limits of the screen, continues to be a fundamental point, as does his ability to simulate within those limits. Among the many merits of Buster Keaton's *Sherlock Jr.*, which he directed and in which he starred as the character, is that of having poked gentle metacinematographic fun at the crossing of the threshold between reality and fiction (which is one of the narrative themes that has constantly recurred since the origins of cinema).

The movie tells the story of a projectionist (played by Keaton) working in a provincial cinema, who falls in love. Accused unjustly by his unscrupulous rival in love of having committed a theft, the young projectionist, who dreams of becoming a detective, tries to clear his name by himself but with disastrous results: His fiancée kicks him out. Disconsolate, he returns to the cinema, falls asleep in the projection cabin, and starts to dream. He dreams of acting in a movie in which he uses his brilliant detective skills to unmask a dangerous criminal. Keaton is, in fact, telling us about the power that cinema exerts over the imagination of the character and, in general, over that of the audience to the point of adjusting dreams, wishes, ambitions, and even concrete behavior as the exhilarating finale shows.

Let us go back to the moment when the sad little projectionist falls asleep (Fig. 2.7). Keaton uses a dimension leaping technique to extract the projectionist's double from his sleeping body, a sort of shadow, a transparent ghost that represents his dreaming self. This shadow evidently does not experience the same reality as the projectionist himself and is thrilled to be able to move freely in the enchanted world of the movie, a dream world in which it is much easier to make sentimental and professional desires come true. On the screen, or more precisely inside the screen, the projectionist will be able to become a detective and win back his fiancée. And now we see Keaton's shadowy double leave the projection cabin, walk down the central corridor, climb up to the screen, and jump inside it. Nothing could be easier—the screen shimmers into mist just like the mirror did in *Alice Through the Looking Glass*. However, it quickly becomes clear that our shadow-projectionist is not completely at ease

Fig. 2.7 *Sherlock Jr.*, Buster Keaton, 1924.

on the other side of the screen; he no sooner gets his bearings in the new environment than something happens to disorient him and he finds himself in danger. The editing! He wants to sit on a stone bench in a garden, but a cut whips the bench out from underneath him, and he lands head over heels in the middle of a busy street in a city; he hardly has time to shake the dust from his clothes and walk along the street with his hands in his pockets when another cut lands him on the brink of a precipice into which he risks falling. Just when he thinks he is safe on the mountain, the scene changes again and he is in the middle of a plain with two lions; then he finds himself on a railway line almost under the wheels of an oncoming train, then on a rock in the middle of the sea, upside down in the snow, and, finally, there he is, tripping over the stone bench in the garden where his adventure began.

It is as if the shadow had kept the corporeality of the "real" projectionist and that this corporeality is irreducible compared to the fragmented space–time of the movie. From our seat in the stalls that space–time seems continuous, transparent, Gorkij would have said artificially uniform, and we would bet our last dollar that we would be able to find our way around it without any trouble at all, but the moment we cross that threshold we would find ourselves in the same situation as Keaton's double. We too would be victims of the clash between the space–temporal continuum of our body and the fragmentary nature of the cinematographic scene. With a sequence that would thrill any surrealist, Keaton tells us that cinema is not life. He pokes fun at a tradition of reflections on the relationship between film and reality that has its roots in the early 1900s, as, for example, that great classic by Edwin Porter, *Uncle Josh at the Moving Picture Show* (1902), which is a brief essay on the impression of reality produced by the moving images and the impalpability of the threshold.[55] However, by revealing its limits, Keaton is also underlining the movie's potentiality for simulation. It is interesting to note that while we watch Keaton's shadow becoming more and more disoriented, we as spectators experience the same feelings of disorientation. The camera doesn't help us, it doesn't give us any shot of reference with the projected images; indeed, the only point of reference is the character, but that is not sufficient for us to understand where we are and in what direction we should go.

But then, in the blink of an eye, our projectionist finds his feet again; the editing doesn't play tricks on him anymore and we, the spectators, are finally able to align ourselves with him, share his feelings, and laugh at the absurd situations in which he ends up. What has changed? Keaton is working on the

[55] Casetti (2008, pp. 145-149).

mediation once more, restoring the intersubjective relationship that he had intentionally disrupted by making us believe that a man in flesh-and-blood could set foot in a movie. He adjusts the visuals, shots, perspectives, camera movements, and the editing so that we can find our bearings again; and here we are, just as if nothing untoward has happened. We are back in our seats. Everything is back in the right place—here in the cinema and there on the screen.

3
Camera Movements and Motor Cognition

> When main-stream audiences around the world pay good money to watch a movie in a darkened theater, they want it to be energized by a moving camera.
>
> Gil Bettman[1]

Style

No matter in which discipline it is studied, the concept of style is always debated; it is elusive and difficult to pinpoint—"unlocalizable" was the term Christian Metz used.[2] In everyday speech style is considered as being different from content: While content has to do with "*what*," style has to do with "*how*." Some years ago, Nelson Goodman stated that although most literary works, fiction and non-fiction alike, have something to say, they usually do other things as well, and we are inclined to think that these other things are questions of style.[3] A few pages further on, he observes that it is often thought that style starts when the facts are finished and feeling takes over, as if style were related exclusively to affective forms of expression rather than to the logical and cognitive elements of a work of art.[4] In the first case style is considered as being able to communicate much of what goes beyond the mere content of the story; in the latter case it is seen as an element that is difficult to define in words, related to emotional spheres that are so complex they have to pass through other channels. John Huston, in his concise but efficacious manner, would say that style is the adaptation of a word or action to an idea,[5] and the challenge that every author faces during a shooting is how to breathe force and visual meaning into an idea that up till then was only expressed in writing or even by word of mouth. In purely pragmatic terms, it can be said that style is everything we can't talk about unless we have actually seen the movie.[6]

[1] Bettman (2013, p. 9).
[2] Metz (1991, p. 183).
[3] Goodman (1975, p. 799).
[4] Goodman (1975, p. 802).
[5] Sarris (1967, p. 264).
[6] Our thanks to Leonardo Gandini for this observation.

The Empathic Screen: Cinema and Neuroscience. Vittorio Gallese and Michele Guerra, Oxford University Press (2020). © Oxford University Press.
DOI: 10.1093/oso/9780198793533.001.0001

There is truth in all these definitions and our daily use of the concept of style, whether applied to works of art or to behavior and lifestyles, is full of these nuances. Style disciplines and strengthens our relationship with artwork (and, in a completely different context, our everyday relationships). Style helps us to get our bearings or even, when circumstances allow, to lose ourselves in fictional worlds. It presupposes a triangulated intersubjective relationship: The artwork with the spectator, and its author/creator with its recipient/beneficiary. When working on the style, there is the possibility of working also on the mediation starting from the type of support used, the equipment/instruments at hand, and the techniques available given the type of equipment.

The etymology of style shows the word has its origins in the Latin *stilus*, the sharp implement the ancients used as a tool for writing; it appeared in English in the early fourteenth century as *stile*. Initially it was related to the instrument used and the techniques dictated by its characteristics, and indeed it is difficult to conceive of style without a tool, just as it is well-nigh impossible to conceive of a tool without the relative processes/techniques required to use it appropriately. David Bordwell, considered style "in its narrowest sense [...] to be a film's systematic and significant use of techniques of the medium."[7] That said, there are few who would attempt to recount the history of style exclusively on the basis of the technologies used and their relative techniques, while many have approached the question of style from the angle of the personality, inspiration, or biography of the artist/author, as well as the historical, socio-economic, and cultural context in which he lived.

Applying this criterion, style identifies not just one single artist or a group of artists sharing the variables specified above, but it also implies the recognition and interpretation of a particular view of the world, often linked to the so-called "spirit of the times." This vision facilitates our understanding of what is behind the immediate appearance of the artwork, its dialectic position vis-à-vis its times, even broadly speaking the artist's role in the society in which he lives and works.

Bordwell drew up the schematic model shown in Figure 3.1 to illustrate the various levels of film cognition.

Generally speaking, most filmmakers conceive style from the lowest level of the model, in other words from the perception of the movie. At this level it is possible to use techniques to modulate what the spectators see, and to attempt to structure their experience on various levels. In other words, this is where comprehension of the movie starts and why filmmakers tend to approach style

---

[7] Bordwell (1998, p. 4).

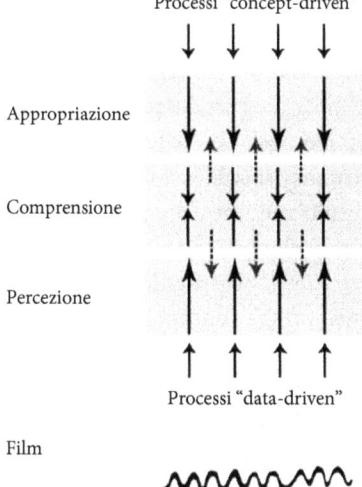

Fig. 3.1 David Bordwell's schematic model of film cognition (2008).
(Concept driven processes/Appropriation/Comprehension/Perception/Data driven processes).
Reproduced from David Bordwell, *Poetics of Cinema*, p. 43, Figure 1.20 © 2007, Taylor & Francis Group.

from a *bottom-up* perspective. However, many film scholars see style as part of the appropriation practices, the cultural relationship that spectators create with the movie from a *top-down* perspective, which obviously contributes greatly to understanding the artwork and its universe.

Regarding the field of film studies, Barry Salt attempted to study the history of film style from the point of view of technological development. He was trying to show that the relationship between the available technologies and film style during a given period of time was direct and binding and that many of the narrative strategies and formal solutions used in filmmaking were dictated by the equipment available.[8] Salt was particularly critical of the position on style held by Bordwell and his colleagues; they held that film style was the result of a socio-economic negotiation and, first and foremost, was related to modes of production in a given social context; for a serious study of the classical Hollywood style, a system of integrated analysis was needed to develop an aesthetic framework starting from the very buildings in which films were projected and their social impact. This framework was to correspond to basic

[8] Salt (1993).

narrative requirements, motivated by full and direct comprehension for the highest possible number of spectators.

According to Bordwell and his colleagues, style is the outcome of formal norms sustained by an integrated production system, whose objective is to establish "how a movie should behave,"[9] which stories can be adapted to which type of medium and how they should be told.

Leaving aside for the moment the flattening effect that the radical adoption of these concepts risks provoking in historical analysis,[10] it is evident that both Bordwell and Salt have contributed significant models for the study of film style. Both have focussed on the concept of mediation, but where Salt limits this to the technological mediation between the spectator and the filmic reality, Bordwell extends it to the enormous ability that cinema has to crystallize components of industrial art in the form of film. The mediation that Bordwell has in mind is similar to Francesco Casetti's concept of negotiation, in which cinema has the ability to give form to the different forces and counterforces of modern times.[11]

Both approaches fall, albeit at different levels, into what Vincenzo Buccheri defined as "intensive stylistics", an idea of style that functions on the top level of film and, in less fortunate cases, could run the risk of interpreting the history of cinema exclusively in terms of changes in technology or statistics linked to one stylistic solution or another.[12] That said, if the objective is to use style to understand cinema on the historical and cultural contexts through which it passes, it is crucial to examine mediation in order to fully understand the relationship (both embodied and not) with film fiction. What Buccheri defined as "intensity" is what filmmakers and producers have worked on for years, prior to any concrete plan of historical reflection which in any case would have been the natural outcome of the effective functioning of narrative and stylistic choices. This is probably what best conveys the idea of the bases of the experiential and cultural changes wrought by film. According to Erwin Panofsky, cinema has the ability to organize things and people "into a composition that receive [sic] its style [...] not so much by an interpretation in the artist's mind as by the actual manipulation of physical objects and recording machinery,"[13] and so is the form of expression that better than any

---

[9] Bordwell et al. (1985, p. XIV).
[10] See Pravadelli on this subject (2007, pp. 31-6).
[11] Casetti (2008).
[12] Buccheri (2010, p. 37 and following).
[13] Panofsky (1995, p. 122).

other renders justice to the materialistic interpretation of the universe that pervades contemporary civilization.

The majority of filmmakers aim at a form of mediation that will involve the spectator's senses as intensely as possible. What appears on the screen, the ways and means, the strategies of appearance in cinema are the result of the direct relationship our perceptual system has with technology and are influenced by production processes and needs that consolidate the social aspects. While in the first case style is used in its etymological sense, in the second it reflects the gradual creation of a canon, stemming from the Greek *kanon* (a cane), indicating a rigid norm.

It is well known that every theory of cinema deals with film style. From the pioneering manuals of the early American theoreticians of the 1910s (of whom we will say more later) to the classical European theory to the great paradigms of semiotics and structuralism or cognitive and phenomenological propositions, the crux is the persuasiveness of the stylistic analysis, which stands out as the true coefficient of specificity. Indeed, the criticisms formulated in any artistic field against the Darwinians or those who sustain evolutionary theories almost always include the accusation of underestimating stylistic themes and specificities of the various media.[14] Giulia Carluccio correctly noted that while the risk of a "generic panstylism" should be avoided at all costs, it is important not to lose the wealth and multidimensionality of stylistic aspects.[15]

Speaking of the "camera's lyricism," Béla Balàzs remarked that it identifies us "not only spatially but also emotionally with the characters in film" and how "the expression of every phenomenon presented to us by the camera set-up corresponds to the impression it makes on other figures within the film."[16] What we wrote in the previous chapter regarding film subjectivity supports this "spatialization of emotivity" that originates from the motor, mimic, and gestural behavior of the camera. The filmmaker uses the style of the movie to adjust the balance between "excess" and "recognition," facilitating the spectator's corporeal involvement at the right moment and giving depth to the opportunities for simulation within the virtual dynamic space. Talking about certain stylistic needs and rhetoric figures to be found in contemporary cinema, Adriano D'Aloia uses the term *"figure tensive"* (tensive forms) to emphasize the

---

[14] Gallese and Guerra (2013a).
[15] Carluccio (2006, p. 139).
[16] Balázs (1931, p. 35).

characteristics of a space "in which the 'fictional' forces burst out of the surface of the screen and pervade the physical and psychic area in which the spectator is seated."[17] The sensory-motor programs that a movie can activate are the result of carefully studied stylistic decisions, that filmmakers initially reached by induction or instinct and considerations regarding the available technical equipment; these programs form the foundations not only for specific stylistic preferences, but also for the very soul of certain concepts of cinema.[18] Clélia Zernik followed a similar line of research and also posed the question as to how to study recurring forms of style used by various filmmakers in the light of what she defined as a "stylistic perception of cinema,"[19] inspired mainly by gestalt and phenomenological paradigms, but also by the attempt to use some contributions from cognitivism.

A study of embodied cognition in film must contemplate the idea that an incorporated continuity exists between our reality and what we see on the screen; to put it succinctly, style serves this kind of continuity (or, sometimes, what programmatically counteracts it).[20] The intersubjective nature of moving images confirms what Vivian Sobchack wrote in her first book on the phenomenology of cinema, in which she says that cinematic language "is grounded in the more original, pragmatic language of embodied existence."[21] If what Art Shimamura says is true,[22] that the way cinema captures the dynamism of human experience is more natural by far than videos made for neuroscientific research, then it is worthwhile trying to understand which levels of resonance are stimulated by given stylistic solutions that are now part of our visual and imaginative habits.

Style in film touches many different aspects, as it does in any art genre: It ranges from the choice of the lens to how the scene is illuminated, from the type of camera movement to the type of editing, from the positioning of the actors in profilmic space to the choice of framing, from the angle of the shots to the acoustic counterpoint, from the richness or poverty of the scene to how dialogue is used and the type of acting required. Some of these elements exclude others, some converge and contribute to the overall effect of the movie. Each and every one of them, however, engages the spectator in a form of confrontation, sometimes easygoing, sometimes challenging.

---

[17] D'Aloia (2013, p. 64).
[18] Reference is made to Cinema 1, the first of two volumes on cinema by Deleuze.
[19] Zernik (2012).
[20] These points have been explored in depth by Guerra (2015c).
[21] Sobchack (1991, p. 13).
[22] Shimamura et al. (2013b, p. 77). See also Shimamura (2013a).

Moving the camera

Many movie-goers would be ready to identify the filmmaker by his use of camera movements, but few would attempt this on the basis of the editing. There are filmmakers who hardly ever move the camera at all, trusting to the implacable and productive duration of the movie rolling out in front of the spectator's eyes. It is true that if there is no (or very little) camera movement, the spectator has difficulty interacting with the movie on the basis of the sensory–motor schemas that are usually brought into play by the style, so weakening the agentive forms. These are compositional decisions that adopt the theatricality of staging to resolve the dialectics between what Victor Freeburg called the "static" and "dynamic" forms, the two aesthetic options that filmmakers must know how to manage and configure.[23] It must be said that while Freeburg, who taught film-making in the first half of the first decade of the 1900s, attributed equal importance to the static and dynamic forms, contemporary manuals decidedly favor the dynamic. Gil Bettman went so far as to say that if a filmmaker opts to exclude or limit camera movement in a movie, he is perfectly within his rights; that said, his final product may appeal to film critics but is unlikely to top the approval ratings with the mainstream audience. According to Bettman, the contemporary visual realm is all for energetic camera movements and audiences are enthusiastic about films that involve them through movement. Aesthetic and intellectual considerations (and there are many) notwithstanding, it must be acknowledged that the theme of visual energy has characterized mainstream cinema from the 1970s onwards and that big-budget filmmakers such as James Cameron, Chris Nolan, Michael Bay, and Peter Jackson, to name just a few, continue to experiment in that direction. As Bettman says, camera movement and the energy it exudes are above all a question of time and money.[24]

We are not saying that a camera that doesn't move doesn't communicate; what we are saying is that the involvement of the average spectator is directly proportional to the intensity of camera movements. In the absence of movement, the editing and arrangement of figures and spaces within a shot can produce a feeling of oppression, which serves to reinforce the message that the movie is attempting to convey. An excellent example of this is Carl Theodor Dreyer's *The Passion of Joan of Arc* (1928), in which the spectator is barraged by a sequence of close-ups expressing the violent shades of power and persecution

---

[23] Freeburg (1918).
[24] Bettman (2013, p. 9).

and the desperate and courageous hues of defeat and sacrifice. There are cases in which the choice of a static style emphasizes the sacredness of the scene and encourages contemplation, stimulating memories that draw on other forms of expression (as in the early films of Pierpaolo Pasolini); there are others in which the still camera obliges the spectator to reflect on his role of observer, abandoning him, alone and vulnerable, to face acts of violence or implacably lengthy scenes of pain (a characteristic of Michael Haneke's cinema); there are others in which the absurdity of existence, reflected in the anempathetic style, is mirrored in the total absence of camera movement, where the spectator finds himself far removed from the action (as, for example, in certain films by Roy Andersson, who won the Golden Lion at the 71st Venice International Film Festival for *A Pigeon Sat on a Branch Reflecting on Existence* in 2014); finally, there are cases in which the lack of camera movement is used as an experimental device to draw attention to the length, theatricality, or materials of the movie.

Therefore, while the absence of camera movement can be taken as a very profound form of style, movie critics usually get much more excited when it is present; from Griffith to Ophüls, from Murnau[25] to Hitchcock, from Kubrick to Bertolucci, just to cite a few, camera movement has become a signature, a sign of virtuosity, of great mastery, but it is also a gesture used to usher the spectator inside the movie and reveal the power of cinema. When the camera moves, its role is not so much that of an observer, but rather that of a participant in the action, both when its participation is anonymous or objective and when it mediates the point of view or the position of one of the characters. That said, as every teacher of film technique knows, the best camera movements are invisible, those of which the spectator is not aware; he should not have the impression that the camera is moving, there should be no flourishes or master strokes; the camera movement should resonate with the movement at low-level so as to perfectly simulate action and emotion. In other words, the ideal camera movement is a movement that does not interfere with the impression of reality that the movie conveys; it must first and foremost serve the purposes of the plot and be drawn by the narrative.

Reflecting on these concepts of camera movement, Robert Zemeckis concluded that there are three types that can be considered invisible: Externally generated camera movements, internally generated camera movements, and moving establishing shots. Bettman called them "Bob's rules" in honour of Zemeckis.[26] A filmmaker who breaks these rules intentionally intends to create

---

[25] See Müller (2014) on Murnau and Lang with regard to these topics.
[26] Bettman (2013, pp. 10–20).

movements that are perceptible, so setting the conditions for a reflection that could be defined, to some extent, as metacinematographic.

We all know that it is possible to be moved by and to resonate with the images of a story and how these same feelings can be produced by the force of a visual and technical solution, which, though acting on a different level, has the potential to produce this same effect.[27]

Returning to the three categories of invisible movement Zemeckis identified, the first category, externally generated camera movements, is the one most commonly used (according to Bettman, 95 percent of camera movements in mainstream movies fall into this category). It is used to follow a character, a vehicle, or an object in motion, which would pass out of the shot and out of our reach if the camera didn't move. The second category, internally generated camera movements, is when the camera follows the line of sight or hearing of the character; this includes not only the point-of-view (POV) shot discussed in the previous chapter, but also those shots in which the camera movement places the spectator in relation with the character's position (either physical or projected) as well as supplying information as to the direction of his gaze. It is important to note that internally generated camera movements can also be used to communicate a character's thoughts or feelings, such as when he suddenly understands what is happening in a specific situation or comes to terms with the psychology of another person and the camera moves in for a close-up with a tracking shot or a zoom. The moving long shots that compose the third category are frequently found at the beginning of a movie or a new sequence, animated by a movement that directs the spectator's attention towards the space in which the story will evolve. Frequently they are a variation of the classic establishing shot that is generally fixed.

Camera movement is therefore both informative and performative. It must in no way violate the transparency of the movie or interfere with the filmic cognition; in fact, it should enhance and strengthen it. In an article on circular camera movements, Lennard Højbjerg stated that certain stylistic features, and particularly those structured around camera movements, enhance the embodied component of our visual perception by adding elicitation of precise feelings and emotions to simple information. Højbjerg holds that the perceptual consequences of circular camera movements are developed at the level of bodily feelings, evoking particular sensations in the spectator that cognitively reintroduce the meaning of the sequence. In fact, a circular camera movement

---

[27] Ed Tan (1996, pp. 64–66) talks of two types of emotion: Those that originate from the fictional world (F emotions) and those that originate from the movie as an artistic artifact (A emotions).

is frequently used in love scenes (an example of this is the meeting between Cinderella and the Prince in Kenneth Branagh's *Cinderella*), even if there is nothing "passionate" or "loving" about the movement itself. According to Højbjerg, apart from the fact that certain movements are replicated movie after movie and therefore settle into our visual unconscious with precise semantic meanings, the advantage of the circular camera movement lies in its embodiment, which is its ability to imbue meaning into bodily patterns that activate metaphoric meanings which are immediately picked up by the spectator.[28] In a case like this, the camera movement is generated by the characters' feelings; in addition, there is the metaphoric contribution that makes it familiar and therefore to a certain extent invisible.

The question of the metaphoric value of camera movements has been dealt with in literature, greatly inspired by the classical works by Eve Sweetser, Mark Turner, George Lakoff, and Mark Johnson.[29] If our ability to produce metaphors is based on the spatialization of abstract concepts and the synesthetic construction of images that call to mind the action of our corporeality on something intangible and elusive, then cinema, which is essentially the spatialization of corporeal actions and intentions, naturally shapes visual metaphors that are based on easily recognizable motor and bodily schemas, often embodied by camera movement.[30] The type of identification that camera movement facilitates (and use of the still camera impedes) is what Turner would call "projected action" regarding the stories that characterize our way of thinking and being; this form of action, as Turner observes, is almost always a "body action."[31]

As it has been constantly pointed out by those who have studied camera movement over the years, the sense of participation in the camera action is undoubtedly enhanced by the fact that its behavior is interpreted by both filmmakers and spectators according to evident and automatic anthropomorphological analogies.[32] The camera moves in three-dimensional space, behaving in much the same way as a human body would in relation to events taking place in profilmic space, which are then exposed on film and rendered into a bi-dimensional form. In a study of the perceptual representation of filmic space through camera movements, David Bordwell underlined the

---

[28] Højbjerg (2014).
[29] Sweetser (1990); Turner (1996); Lakoff and Johnson (1980).
[30] Marìa J. Ortiz (2014) proposes an interpretation of certain base elements of the cinematographic mis-en-scène, starting from the concept of the primary metaphor. See also Coëgnarts and Kravanja (2012, 2014).
[31] Turner (1996, p. 38).
[32] See, for example, Branigan (2006, p. 36 and following) and Nielsen (2007, p. 14).

importance of grasping these forms of bodily contact, which are not restricted simply to the camera/head/gaze equation but entail possibilities of movement that mime the behavior of a complete organism, thus giving the spectator a sense of kinetic and haptic depth regarding the spaces crossed and the objects touched.[33] From this point, movement can access metaphorical meanings that give rise to shared cinematographic semantics of bodily origins. In the 1930s Rudolf Arnheim explained that thanks to this possibility of movement the "mobile camera" could provoke "feelings of giddiness, vertigo, intoxication, falling, rising" but above all that "the film artist is thereby able to do what is hard for the theater director, namely to show the world from the standpoint of an individual, to take man as the center of his cosmos."[34] The "man" Arnheim is talking about is none other than the camera.

Camera movements take place on the set and have to do with a real object moved by a human being according to the instructions received from the filmmaker. The body of the person operating the camera is part of the narrative incorporation strategies that are revealed in the camera movements. Film is capable of reproducing the movements of the human body, its physicality, forms of attention, and conflicts of impulse that keep us suspended between balance and imbalance. Through the camera, the body of the cameraman challenges the profilmic space that will eventually end up, quite transfigured, on the screen.[35] The Steadicam has greatly contributed to intensifying this relationship; Garrett Brown, its inventor, said that ideally we should think of the camera as Margot Fonteyn and the cameraman as Rudolf Nureyev, bound together in a dance in which the camera transmits the cameraman's expressiveness in three dimensions; as the cameraman gradually becomes more skilfull, the camera is capable of going with him almost everywhere and does everything he imagines doing.[36]

There is still very little literature available on camera movements, always considered as too elusive from the analytic point of view, but Sobchack wrote an important essay on the subject almost 30 years ago in which she applied Merleau-Pontian phenomenology[37] to the question, laying the foundations for a theoretical discourse that inspired the experiment on camera movements that we will expound in the final section of this chapter. She started from the assumption that however far the camera can go beyond our visual or physical

---

[33] Bordwell (1977, p. 19–25).
[34] Arnheim (1957, pp. 111–12). See also Gallese and Guerra (2015).
[35] MacDougall (2006, p. 3).
[36] Brown, in Ferrara (2001, p. 7).
[37] Sobchack (1982).

possibilities, it will still be perceived primarily as being able to express and to live space in a human, as opposed to mechanical, fashion. For this reason, the "semiotic structure" of camera movement is based on two essential concepts for phenomenology, i.e., embodiment, and intentionality, as conceived in Merleau-Ponty's philosophy.

Most of the time we are oblivious to the fact that the camera has moved; the movement is introjected pre-reflexively to a certain extent without actually reaching the experiential level of our consciousness; in fact, Sobchack says that we experience most camera movements at the same level of consciousness with which we experience our own body. Of course, there are instances in which the complexity of the movement, the fact that it cannot be reduced to our motor or visual potential, renders it "opaque"; in other words, the transparency that should allow us to understand it unconsciously is eliminated (these are the instances that violate "Bob's rules"). From this perspective the camera is a "corporeal subject" that sees, moves, and expresses a perception. At this first level,

> The moving camera is not only a mechanical instrument, an object of visual and kinetic perception; it is also a subject that sees and moves and expresses perception. It participates in the consciousness of its own animate, intentional, and embodied existence in the world. Thus, at this primary level of the body subject, the viewer intersubjectively and prereflectively recognizes and understands the camera as sharing the manner of his or her own existence, as manifesting the material and kinetic code of an embodied and intentional consciousness.[38]

On the basis of Merleau-Ponty's definition of consciousness, Sobchack observes that our intimate and carnal knowledge of the camera has more to do with "I can" than with "I think" and how the camera's "I can" is achieved by moving around the "natural space of the screen,"[39] in living in that space and transforming it into an anthropological space, creating the ideal experiential ground for the successive abstract reflections that will fall into the pertinence of its "I think".

The next step should be a research study to help understand the forms and levels of resonance elicited by cinema's four fundamental movements: Those of the characters and objects present in the shots; those *between* shots, i.e., the editing; the optical movements produced by devices such as the zoom; the

---

[38] Sobchack (1982, p. 327).
[39] Sobchack (1982, p. 318).

physical movement of the camera itself (the movement that Sobchack purposely defined as "bodily motion of the camera").[40] We will go on to examine the latter two types of movement in this chapter and will deal with the second type in the next chapter. It is fascinating to see that the human motor cortex retains and modulates the intimate nature of movement even when the rules that make it invisible are applied. Before going further, we will take a break with Stanley Kubrick—a justified break as what we will see here will come in useful later on when we talk of zooms, dollies, and Steadicams.

Kubrickian Intermezzo

The greatness of Stanley Kubrick lies in how he was able to create movies in which clarity and narrative involvement combined perfectly with technical research on the forms of film that never fail to fascinate and enthral the spectator. Kubrick was one of the few filmmakers, who grew up in the days of classic cinema, to be still widely acclaimed by the new generations; this is to be attributed to the fact that his cinema is a combination of the great Hollywood narrative tradition, rigorous experimentation, and an authorial independence in European-style that still managed to adapt perfectly to the American productive structures and developed a coherent dialogue embodied in two major themes, violence and power. In a famous interview conceded to *Playboy* in 1968 (the year that *2001: A Space Odyss*ey debuted), Kubrick stated that above all he tried to "create a visual experience, one that bypasses verbalized pigeonholing and directly penetrates the subconscious with an emotional and philosophic content."[41] Kubrick was very clear about how cinema could speak to the senses of the spectator. His perfectionism, his formal manias, and the almost paranoiac care he took with the quality of the projection of his movies, were part of a philosophy of cinema and a concept of authorship that ran parallel to his main themes, to his unique and unforgettable characters and choice of subject; Kubrick gave the same importance to all these aspects.

Many moviegoers would be able to recognize the Kubrick touch, his view applied to film, but as Marcello Walter Bruno observed some years ago, there

---

[40] Although she does not make a direct reference to Arnheim, Sobchack here supports the subdivision he proposed: (1) The movements of objects, alive or dead, that are photographed by the camera; (2) the effect of the perspective and the distance of the camera from the object; (3) the effect of the moving camera; (4) the synthesis of individual scenes, accomplished by montage, in an overall composition of motion; (5) the interaction of movements placed next to each other (Arnheim, 1957, pp. 181–2).
[41] Norden (1968, p. 85).

are no real "Kubrickian" movies as such.[42] His cinematographic approach is so masterful, so pure and absolute that there are no successors or imitators. He was continuously changing genre, range, and, consequently, style. His every movie was a test to radicalize themes and canons: Film noir, war movies, black comedy, science fiction, historical movies, horror, and even melodrama. And each time we watch one of his movies, we have the feeling that Kubrick has had the last word, that he has created the movie to end all movies, a cult movie, whether the genre be science fiction, historical, or horror par excellence. At the same time, however, each movie is a technical test, in which he challenges and exploits to the maximum the photographic potential (Kubrick started his career as a photographer) and the dynamic and audiovisual possibilities of cinema. His proclivity for change and experimenting new forms is undoubtedly attributable to his inexorable research in the technical field; after the extraordinary achievement of *2001* (which was a turning point for this aspect), the movies that followed seemed to be constructed to test the resistance of a given style, taking over the visual atmosphere of the movie, stunning our senses, and remaining anchored in our memory.

His technical virtuosity ranges through *A Clockwork Orange* (1971), in which the use of a wide-angle lens greatly taxed camera movements, still camera shots, and slow motion; *Barry Lyndon* (1975), the movie of the zoom; *The Shining* (1980), the Steadicam movie; and *Full Metal Jacket* (1987) in which he alternated the use of the dolly and hand-held cameras. This decision to make the tool, the *stilus*, the foundation on which to build the visual and emotional elements of the movie, this determination to transform the theme of the movie into a technical challenge that regulates the interaction between the spectator and the screen, firmly positions Kubrick's cinema narrative as avant-garde.

His use of the zoom in *Barry Lyndon* has the effect of distancing the spectator from the pictorial world of the plot so that Barry's story is continuously "cooled down," an effect achieved also by the anticipations given by the voice-over and the dissociation of the spectator from what he thought was the heart of the plot.[43] Alexander Walker defines this choice as sensual; he lingers on the estranging and meditative role of the zoom and on how few dramatic close-ups or immersive camera movements are used; he remarks that the effect is rather like being on a tour of an art gallery with a guide who keeps telling you not to go up too close to the paintings.[44] Enrico Ghezzi went so far as to see in the use of

---

[42] Bruno (1999, p. 108).
[43] Guerra (2019).
[44] Walker (1971/1999, pp. 243–5).

the zoom the heart of the dialogue on history that Kubrick developed in *Barry Lyndon*: The false movement, by creating an illusory effect of center, suspends all possibility of judgment and modifies space without actually covering it.[45] The zoom in *Barry Lyndon* leaves the spectator with the sensation that there is something regulating the space and tempo of the movie that neither the characters nor the spectator can control. There is no sharing of agency in movement (which, in fact, is *not* movement), the diegesis is violated by the zoom that seems to tell us, time and time again, that Barry (and mankind in general) is by no means the principal character of the scene. The zoom-out is used more often than the zoom-in, with the clear intention of showing that Barry's destiny cannot be controlled and that he is guided by a somewhat mysterious and irrational force.

This same force, however, is implied by the zoom-in, for example in Lady Lyndon's superb entrance. In this case the spectator is aware that he is being taken somewhere, that there is something on which he has to focus his attention; while in the zoom-out the iconic annuls the diegetic, in the zoom-in the reasoning of the diegetic limits that of the iconic. Lady Lyndon's first appearance on the scene is one of the most elegant sequences of the movie; to the notes of Schubert's Piano Trio, a single, complex shot takes the spectator onto a terrace with an initial frontal movement, then swings laterally to show a group of noblemen seated at tables, perfectly immobile, as if in the scene of a painting. The dolly stops at the table at which Barry is seated and shows a group of four people moving in the background, following the geometry of the flowerbeds. As always in the movie, the voice-over hastens to remove any possible element of surprise, informing the spectator that Barry is thinking of marrying money; at that point Kubrick uses a deep zoom-in on the four figures of the family of sir Charles Lyndon, flattening them, maintaining the focus on Lady Lyndon herself (Fig. 3.2).

The outward effect of the movement, its "unshareability," indicates that the character (and the spectator with him) has no option but to look and also that his gaze is locked to the optical motion of the zoom.

The feeling that there is a phantasmal presence that makes the use of the zoom so disturbing in *Barry Lyndon*, is more evident in *The Shining*. The impression that Overlook Hotel is a living organism, still pulsing with the mysterious and violent events entrapped in its foundations, in its rooms and corridors, becomes tangible in the shots in which it seems that someone is

---

[45] Ghezzi (1995, p. 128). On the "irruption" of various camera moves and the role of the zoom as communicating an ineluctable destiny, see Guerra (2007, pp. 91, 104–5).

Fig. 3.2 *Barry Lyndon* (Stanley Kubrick, 1976).

walking or running behind Danny or Jack Torrance. Here the camera movement is generated by external factors, drawn by the child's tricycle, or Jack's body in the labyrinth or his terrified wife Wendy, but the spectator has the impression that it is generated by internal factors, that the shot is a POV shot by an unknown entity, maybe one of those mysterious presences that haunt the hotel. Once again Kubrick needed a recurring and persuasive filmic gesture to accompany the narrative; like Delmer Daves when he was shooting *Dark Passage*, Kubrick needed a camera that would move like a human being even if in his case the corporeality to be expressed was phantasmal. The solution he found this time was technical: The Steadicam. Invented by Garrett Brown in the mid-1970s, this device is strapped on to the cameraman rather like a piece of armour and is equipped with a sophisticated system of buffering that prevents any form of vibration or unsteadiness; it performs well even under particularly difficult conditions such as sequences in which the cameraman has to run. From the very beginning Garrett Brown declared that the main objective of the Steadicam was to simulate human vision to perfection,[46] leveraging the sense of stability that characterizes our perception of reality (fortunately, we are not able to perceive all the micro-movements and changes in direction of our head; we see the world from a stable position).

[46] Brown, in Ferrara (2001, p. 104).

Larry McConkey, one of the world's best Steadicam operators, has said that by using this device "the camera becomes like another person and the audience becomes connected through that person to other actors. The audience become more empathetic, more involved."[47] The Steadicam perfects the perception of being of a virtual body, intensifying the spectator's feeling of being at the center of the story and facilitating the simulation of the interaction with the spatial element of the movie and with its characters.

This, of course, was absolutely ideal for Kubrick, who needed to build this dense and terrifying web of relationships. Talking to Michel Ciment during an interview, Kubrick explained,

> The Steadicam allows one man to move the camera any place he can walk—into small spaces where a dolly won't fit, and up and down staircases. [...] You can walk or run with the camera, and the Steadicam smooths out any unsteadiness. It's like a magic carpet. The fast, flowing camera movements in the maze would have been impossible to do without the Steadicam. You couldn't lay down dolly tracks without the camera seeing them and, in any case, a dolly couldn't go around the right-angled corners of the maze pathways. Without a Steadicam you could have done your best with the normal hand-held camera but the running movements would have made it extremely unsteady.[48]

The cinematographer John Alcott, who worked with Kubrick on the sets of *2001, A Clockwork Orange, Barry Lyndon*, and *The Shining*, said that Kubrick was fascinated with the hand-held camera and when this device was being used as in some of the best-known sequences in *A Clockwork Orange*, he liked to do the shooting himself in order to get the camera's relation with space and the actors absolutely right. Alcott tells us that Kubrick had found a way to hold his Arriflex that guaranteed an incredible degree of stability, to the extent that it could have qualified as a "Steadicam *ante litteram*."[49] It could be said that from the very beginning, or at least from the time of *Paths of Glory* (1957), Kubrick was concerned with the problem of stabilizing the camera and of how to simulate the presence of a human (or phantom) body in his movies. But a closer look at some of the better known and now-legendary sequences in *The Shining*, in particular the scenes with Danny's tricycle, shows the enormous potential of the Steadicam (Figs. 3.3 and 3.4).

[47] Brown, in Ferrara (2001, pp. 123–4).
[48] Ciment (1980).
[49] Ciment (1980, p. 220).

Fig. 3.3 Danny on his tricycle (*The Shining*, Stanley Kubrick, 1981).

The idea of having Danny race through the rooms and corridors of Overlook Hotel on his tricycle is one of the movie's winning touches, rendering to perfection the impression that the hotel and its inmates are haunted by disturbing presences. The spatial–temporal short circuit of the plot solidifies in two places, the hotel and the labyrinth, and the presence of the Steadicam welds together two dimensions, those of the Torrance family and the mysterious presences hiding in the depths of the hotel. It is no accident that at the end of the movie the camera movement remains alone, unrelated, in the interior of the hotel, abandoning us to the mystery of the photographs from the 1920s which record a young Jack at a ball in the hotel on July 4, 1921.

KUBRICKIAN INTERMEZZO 103

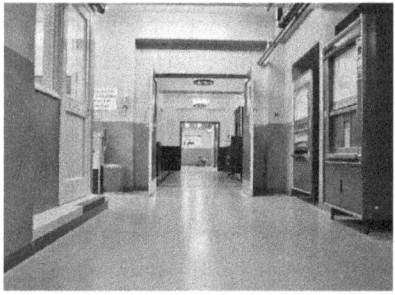

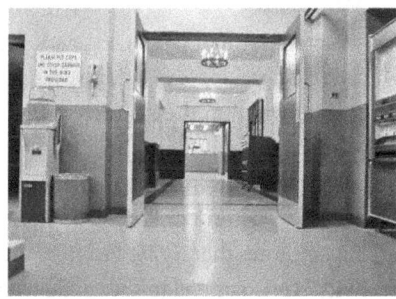

Fig. 3.4 The long shot in the kitchen corridors (*The Shining*, Stanley Kubrick, 1981).

The function of the Steadicam is seen to best effect in these tricycle scenes; the cameraman can literally run behind the child and the stability and fluidity of the end result are astounding; if we ourselves were to run behind the tricycle we would perceive the scene in much the same way (true, the height would be different, but as has already been seen when analysing *The Spiral Staircase*, simulating movement gains more from stability and fluidity than from the assumed height of the observer).

But there is a shot where it is clear that Kubrick has gone to great lengths to make the spectator aware of a presence, human or ghostly, behind the child: The shot in which Danny crosses the wide spaces of the corridors of the hotel's kitchens. This time the Steadicam operator is not shadowing Danny closely, he is keeping his distance from the tricycle, and the speed at which he moves is different to that of Danny's vehicle. The lack of correspondence between the speed at which Danny is pedalling and the camera movement radically changes the spectator's perception of movement; it is no longer externally generated, it has become internally generated. In other words, the presence of another being in the scene, embodied in the camera movement, is sharpened and made more convincing by the Steadicam's visual–kinetic potentialities. As Danny moves further away into the long shot the footsteps of the Steadicam operator gradually become more audible; when the child leaves the scene altogether, contrary to what would normally happen there is no cut to another scene, the spectator is left with the ominous movement of the "phantom" moving forward. The fact that there is no physical body to be seen does not have any adverse effect on the continuation of the camera movement, which, on the one hand, indicates that it has its own autonomy and therefore is to be attributed to another agent, and on the other, that Kubrick is relying on our resonance with a camera movement in an empty space.

Ruggero Eugeni has pointed out that of the classic Kubrick elements identified by various movie critics, corridors (whether real or metaphorical) function as a sort of vortex, swallowing up the characters in the scene.[50] He went on to observe that this same role should also be attributed to the subjectivity or semi-subjectivity of which Kubrick makes extensive use in all his films and which is connected to precise stylistic decisions such as the use of the tracking shot or the Steadicam. Eugeni notes that in *The Shining* "the gradual expansion of the regime of subjectivity" runs parallel to the gradual descent of both characters and spectators into a state of hallucination and uncertainty as to what is true. This falsified perception/projection is emphasized by the motor element that the Steadicam can enhance more effectively than any other technique, and that intensifies and fine-tunes the levels of resonance and substantiates the regime of subjectivity.

In the films he made after *The Shining*, Kubrick continued to respect his commitment to incorporate a sense of film in cinematographic techniques. *Full Metal Jacket* is divided into two parts; the first is set in the Parris Island marine boot camp, the second in Vietnam, during the war. Kubrick works on techniques of which he has already made wide use in his earlier films: Tracking shots in all their potential and the hand-held camera; the former being used to

---

[50] Eugeni (1995, p. 174).

highlight the physical and psychological training of the marines, and the latter plunging the spectator into the hell of Vietnam.

Going into more detail, the first part of the movie narrates the foul-mouthed Sergeant Hartman's violent reprogramming of the recruits' minds and bodies, reflected in maniacal cleanliness of the dormitory, the geometrical precision of the bunks, and the perfect lines formed by the soldiers in T-shirts and underpants for the sergeant's control. On Parris Island they kill the man and then resurrect him as a (war) machine. Kubrick dwells on this aspect, mechanizing the movements that characterize the first part of the movie to emphasize the artificiality of the project, his view regarding the annihilation of the personality and the reprogramming of the individual. Private Gomer Pyle is the fly in the ointment, the recruit who cannot be reprogrammed and who finally blows his brains out, not before having rendered the same service to his nemesis Hartman. Gomer Pyle breaks the symmetry: He is too tall, too fat, too clumsy (he is always dropping things, moving when he should be still). He is the "uncooperative body," who cannot cope with the training, but also interferes with Kubrick's aesthetic claim to mechanize Parris Island; Kubrick conveys this to the spectator by breaking the horizontality that characterizes this part of the movie, and films him from above, sometimes moving in on him with the zoom, using the dolly for choice.

And then there is the tracking shot. As in other films, the key to Parris Island is embodied in the choice of technique and here Kubrick opted for the tracking shot. The clean lines of the camera's path, restricted to the tracks, produces a precise movement with no risk of swerving to one side or the other and for this very reason conveys an impression of something unhuman in spite of the fact that the boot camp is populated by a group of young men who interact constantly and in a more or less natural manner. As Michel Ciment observed, in Parris Island the arrangement of the places, the characters, and naturally the use of the camera are perfectly functional to convey the rigour of the training and "suggest an implacable, oppressive and claustrophobic logic" (Fig. 3.5).[51]

Moving mirrors

When we first considered conducting an experiment to measure the impact of camera movements on the brain of the spectator, we took as a starting point the idea that each different type of movement by which a camera films an action

[51] Ciment (1980, p. 244).

Fig. 3.5 The dormitory on Parris Island (*Full Metal Jacket*, Stanley Kubrick, 1987). According to Georg Seesslen, "the camera 'imitates' the military vision, in the first place the absurd order in the training camp—rarely was Kubrick's central camera position so cool and so exact—and in the second place the absurd disorder during combat." What we will see now is how the choice of the tracking shot implies a different concept of spectator and levels of resonance compared to the zoom and the Steadicam.

implies a particular type of physical relationship between what appears on the screen and the person observing it. According to our hypothesis, these specific relationships are essentially motor related; the presence of the camera and how it moves induces different responses from the embodied simulation mechanisms produced by the activation of the mirror neurons in the observer's brain.[52,53]

As explained in the first chapter, mirror neurons are motor neurons present in a cortical circuit connecting frontal and posterior parietal areas; they activate during both the execution and observation of actions and movements, enabling a particular, more direct—we could even say internal—comprehension of the motor goals and intentions that characterize many aspects of the behavior we observe in others. By simulating other people's motor behavior, we are able to understand it from within, reusing part of those neuronal resources that we normally draw upon when we ourselves carry out the same movements or actions. Experiments on mirror neurons in human beings mostly use films showing people in the act of grasping objects or participating in activities such as dancing. These experiments have one aspect in common, the subjects are filmed with a still camera. In real life, however, we are often moving when we observe the world around us; this is why it is necessary to study to what extent and how the observer's movements are able to adjust the response of the mirror neurons.

---

[52] Seesslen (2004, p. 211).
[53] Heimann et al. (2014).

The purpose of this research is twofold, as is the case for many other possible studies on relationship between cinema, body and brain. On the one hand, it promotes our understanding of how the physical approach to a scene of daily life can influence our ability to perceive it and the sensations it transmits to us; on the other, it explains how cinema has developed techniques and fine-tuned stylistic approaches that mimic this ability and increase the effect of the simulation processes. In the first case, cinema (in our specific case, the camera and its repertoire of movements) is instrumental in recreating perceptual conditions in a laboratory that are ideal for reproducing the movements that an observer might make in relation to a given scene. By alternating films showing someone doing something in different ways, the participant (who is sitting still, in front of the screen) is presented with a perceptual situation reproducing some of the ways in which he would normally observe the world about him. For example, when walking around a city, we often see something or someone doing something that attracts our attention and we stop to watch. Or we might move towards someone or something, and while we move the contours and details of the person or the object become clearer as we get closer until we are able to distinguish the identity of who or what we are looking at.

Cinema can reproduce these conditions up to a certain point, allowing us to empathize with the story narrated by the images we are watching on the screen or, as is now more often the case, on a monitor (tablet, computer, smartphone). In our study, the moving images are the perceptual stimulus in an experiment designed to help understand how the position of the observer while observing someone doing something influences the content of the vision that is obtained. Now, as we have seen in the first chapter, if this were the only purpose of the study, cinema would simply be instrumental in helping us understand how we build our image of reality.[54]

However, if the approach is fully committed to experimental aesthetics, it has a second objective, particularly important in the research we are about to describe, consisting of trying to understand how our corporeality underpins our response to various cinematographic techniques. In the case in point, the question is in what does the spectator's response to the various ways of using and moving the camera consist? This moves the issue squarely into the cinematographic and aesthetic field, where neuroscience is merely instrumental for studying the type of effect on the spectator. High-density

---

[54] This brings up the theme of the relationship between cinema, subjectivity, and the sense of reality. Cinema contributes to building the spectator's subjectivity, conditioning his perception of what is considered as real. This will be discussed in depth in the last chapter.

electroencephalogram (EEG) studies performed on the brains of people watching films help to improve our understanding of how and why different filmic techniques have different effects. In our view, the added value of this approach is that by interacting on common projects, neuroscience and the study of film will reciprocally enrich their own research, given the use of methodologies that give greater substance not only to the hypotheses linked to the motor energy deployed during the filmic experience, but also, potentially, to any other psychological themes that those who study film decide to investigate.[55]

This dual objective has guided our approach; we conducted an experiment using high-density EEG to study the cerebral responses of a group of spectators watching the same scene involving an actor or actress standing in front of a table, intent on grasping an object lying on its surface. The scene was filmed using four different techniques: (1) With a still camera positioned at a distance of 2.60 meters from the scene; (2) adjusting the lens of the camera to zoom in on the action; (3) using a dolly, moving the camera closer to the scene along fixed tracks; (4) walking towards the scene using a Steadicam, so that the camera reproduced the movement of the operator (Fig. 3.6). In conditions (3) and (4) the camera was stopped 80 cm from the scene.

The objective of the experiment was to find the answers to two questions: (1) Does the mirror neuron mechanism respond differently when observing a film of a person grasping an object shot with a still camera or a camera reproducing movement with different techniques? (2) Does the mirror neuron mechanism respond differently to different ways of moving the camera? The experiment also investigated whether the differences between the techniques could be correlated to the responses the participants gave at the end of the experiment regarding the feeling of involvement with the scenes and the degree of naturalness or artificiality of the camera movements used in the clips.

The experiment was conducted in two parts. In the first, the EEG responses were recorded while the participants watched the videoclips; in the second part, they were shown the same clips again, then were asked the following questions: (1) To what degree did you feel involved with the scene you have just seen? 2) To what extent did you identify with the actor/actress? (3) How close did you feel you came to the scene? (4) How comfortable did you feel while observing the scene? (5) How realistic did the camera movement seem to you? (6) To what extent did you feel that the camera movement resembled that of a

---

[55] Adriano D'Aloia and Ruggero Eugeni also took a step in this direction when they chose *Neurofilmology: Film Studies and the Challenge of Neuroscience* as the title of the 2014 edition of *Cinéma et Cie* on film and neuroscience.

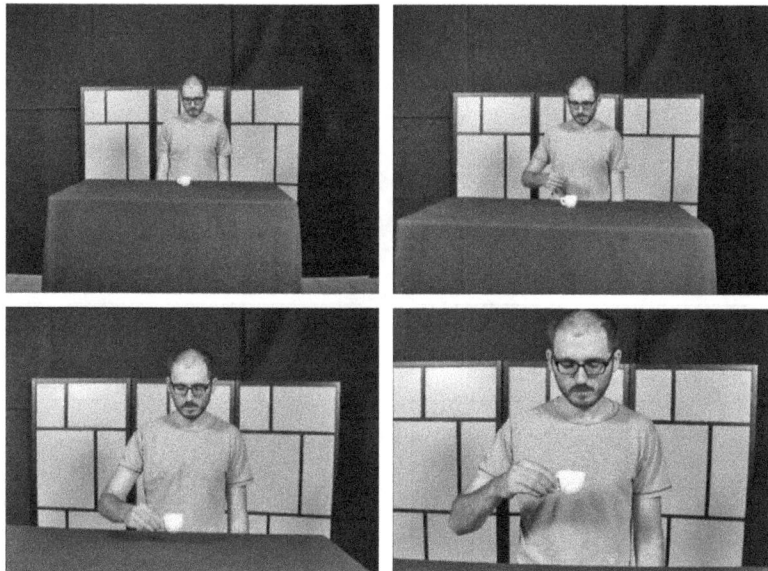

Fig. 3.6 Stills from a video clip in which the Steadicam moves towards the actor.
Reproduced from Katrin Heimann, Maria Alessandra Umiltà, Michele Guerra, and Vittorio Gallese, "Moving Mirrors: A High-density EEG Study Investigating the Effect of Camera Movements on Motor Cortex Activation during Action Observation," *Journal of Cognitive Neuroscience*, 26:9 (September, 2014), pp. 2087–2101. © 2014 by the Massachusetts Institute of Technology, published by the MIT Press.

person moving towards the scene? (Questions 3, 5, and 6 were not asked following the viewing of the video clip shot with the still camera.)

Figure 3.7 illustrates the experimental paradigm of the first part of the experiment (the EEG responses). Each trial started with a fixation cross positioned in the center of the screen, followed by a video showing an actor or actress standing in front of a table, grasping an object (a cup, ball, glass, roll of adhesive tape, etc.) lying on it. In 20 percent of the trials, this clip was followed by the image of an object. The participants had to indicate whether this object was the same as that shown in the video clip by clicking the mouse with the index finger of their right hand. These trials were used to maintain the participants' attention and also to record cerebral activity while performing an action (the clicking of the mouse). The video clip (or the image of the object, if used) was followed by a dark-gray screen. The experiment consisted of 400 trials subdivided in five blocks, separated by a short interval. Each block comprised 80 video clips, 16 for each of the five conditions (still camera, zoom, dolly, Steadicam, motor response from the participants).

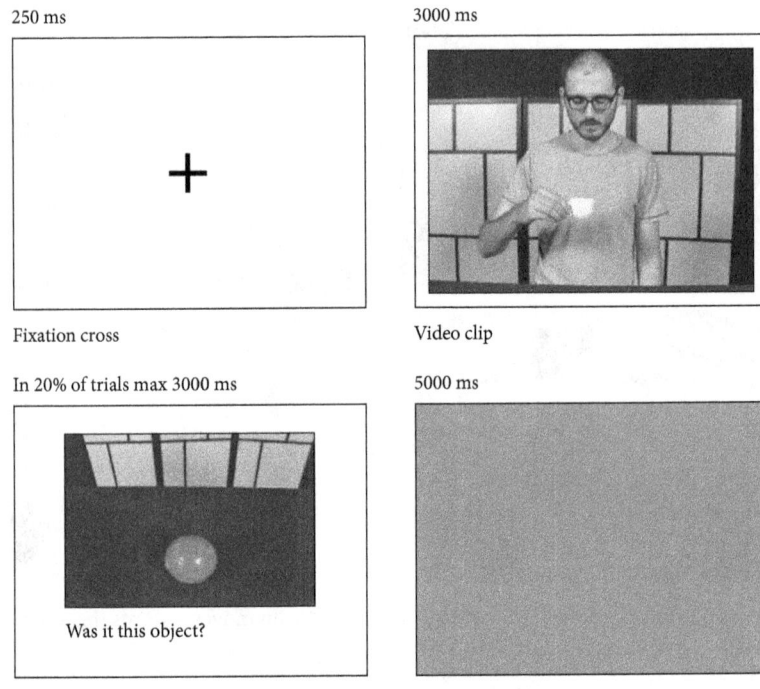

Fig. 3.7 The experimental paradigm during the electroencephalogram recording. Fixation cross (250 ms); video clip (3000 ms); in 20 percent of the 3000 ms "Was this the object in the scene?" Attentional test; gray screen to record the resynchronization (5000 ms).

Reproduced from Katrin Heimann, Maria Alessandra Umiltà, Michele Guerra, and Vittorio Gallese, "Moving Mirrors: A High-density EEG Study Investigating the Effect of Camera Movements on Motor Cortex Activation during Action Observation," *Journal of Cognitive Neuroscience*, 26:9 (September, 2014), pp. 2087–2101. © 2014 by the Massachusetts Institute of Technology, published by the MIT Press.

The EEG experiment enabled us to record the event-related desynchronization (ERD) of the mu rhythm and so document the activation of the motor cortex of the participants. Mu rhythm constitutes a standard marker of the activation of mirror neurons during the performance and observation of actions. Using EEG it is possible to record complex electrical activity produced by the neurons in the brain. The action potential (the code the neurons use to "communicate") oscillates at a series of frequencies or rhythms, including the alpha/mu (8–14 Hz) and beta/mu (14–20 Hz). When the neurons activate, increasing the firing frequency of their action potentials, the frequencies recorded by the

EEG electrodes applied to the scalp of the participants are desynchronized. This desynchronization is recorded by certain electrodes and not by others, meaning that the activation of a particular group of neurons is activated according to the task at hand.

It has been known for many years that the desynchronization of the alpha and beta rhythms takes place in the motor regions of the brain, registered by the central electrodes (C3–C4) every time an action is performed, but recently it was discovered that it also occurs when we observe the action being performed by someone else and not just when we perform it ourselves. Therefore, the desynchronization of the alpha and beta rhythms recorded by the central electrodes is equivalent to recording the activation of the mirror neuron mechanism. The motor nature of this activation is demonstrated by the absence of any significant change in the rhythms, which would be picked up by the electrodes positioned to the rear in correspondence with the occipital lobe where many visual cortical areas are located.

As mentioned earlier, the purpose of this experiment was to verify whether the observation of the same scene filmed with four different techniques (still camera, zoom, dolly, and Steadicam) would evoke a different response from the mirror neurons, influencing the ERD of the alpha and beta rhythms in the brains of the participants.

The results were positive. Shortening the distance between the participant and the scene by moving the camera closer to the actor or actress resulted in a stronger activation of the motor simulation mechanism expressed by the mirror neurons. As was to be expected, the desynchronization of the alpha and beta rhythms at the lowest frequency (14–20 Hz) was significantly more intense when the scene was filmed moving the camera than when the still camera was used, while of the three camera movement techniques, the Steadicam was most efficacious in evoking the activation of the mirror neuron mechanism (Fig. 3.8). The scenes filmed with the zoom evoked the least intensity of activation of the motor cortex, and this was confirmed by the strongest event-related synchronization (ERS) of the beta rhythm (a more rapid extinction of the response) during observation of the scene.

There was no significant change in the recordings by the occipital electrodes positioned in correspondence with the visual cerebral areas, indicating that the responses returned by the motor cortex for the various camera movement techniques did not depend on a non-specific greater attentional response to the stimuli filmed by the camera in movement.

The results show that in this experiment the participants found that the Steadicam most closely reproduced the effect of someone really walking

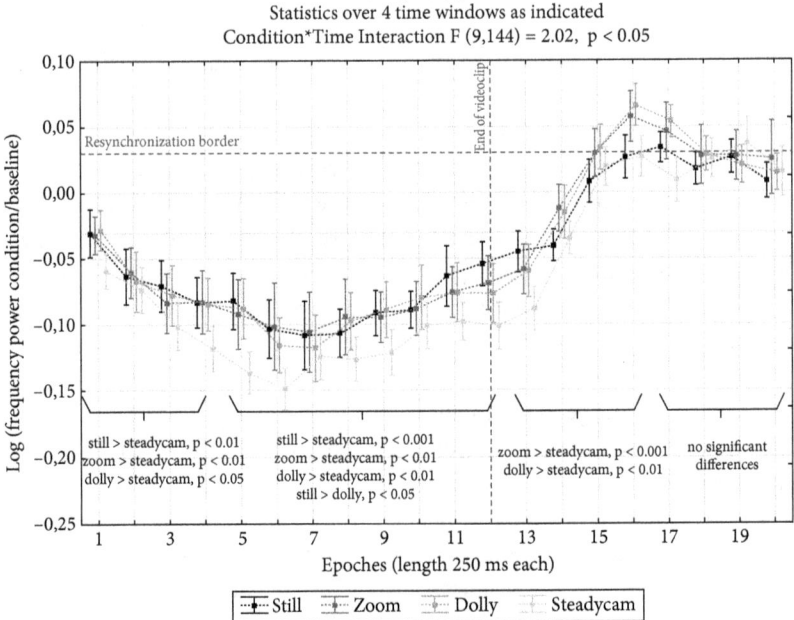

Fig. 3.8 Event-related desynchronization (ERD) and event-related synchronization (ERS) of the low-frequency beta rhythms (14–20 Hz) recorded by the central electrodes at 20 epochs while the participant was watching the video clips. Statistically significant differences ($p<0.05$) were found in four epochs. In the graph, from left: log (power frequency for each condition/base activity); still camera; ERD limit; end of video clip; no significant difference. Epochs: duration of each period = 250 ms.

Reproduced from Katrin Heimann, Maria Alessandra Umiltà, Michele Guerra, and Vittorio Gallese, "Moving Mirrors: A High-density EEG Study Investigating the Effect of Camera Movements on Motor Cortex Activation during Action Observation," *Journal of Cognitive Neuroscience*, 26:9 (September, 2014), pp. 2087–2101. © 2014 by the Massachusetts Institute of Technology, published by the MIT Press.

towards the scene. They also perceived the Steadicam movements as being the most natural and therefore having the highest potential of evoking the sensation of walking towards the scene (Fig. 3.9).

In synthesis, the results of this study show that the filming technique predicts the temporal trend of the activation and return to the rest condition of the participants' motor cortices. Of the video clips that showed a closing of the distance between the participant and the scene, those which simulated a natural view of someone walking toward the scene evoked a stronger motor simulation compared to those filmed with a still camera. Moreover, the stiltedness of the

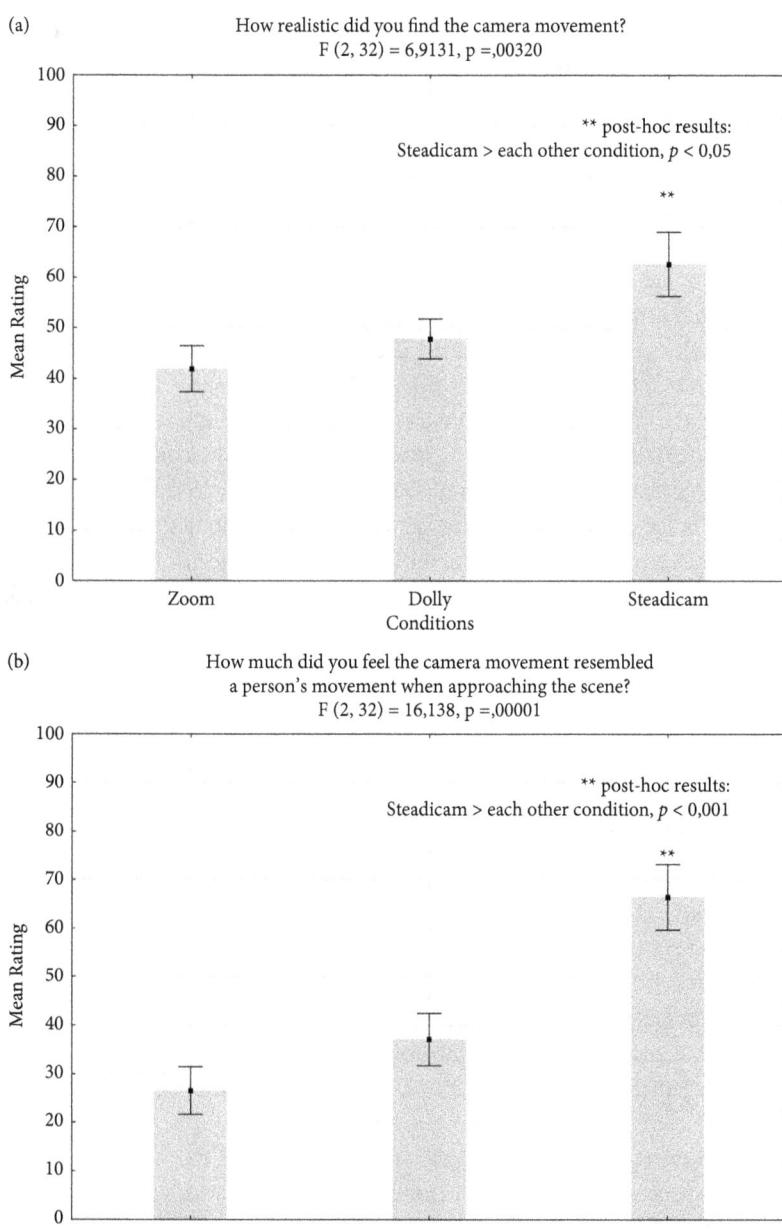

Fig. 3.9 Results of the participants' evaluation of the video clips, in answer to the question: "How realistic did you find the camera movements?" **Post-hoc test results: Steadicam > any other condition, $p<0.05$.

other methods of simulating the dynamic closing of the space (the zoom and the dolly) emerged in the different temporal trend of the ERD and ERS phases of the alpha and beta rhythms registered by the participants' motor cortices. This temporal trend is modulated by the resemblance between motion in the scene as produced by the effective camera movements and as perceived in the real world. In this case the congruence between the perception of movement produced by the Steadicam and perception of real movement results in the maximum intensity of motor simulation.

This study has produced the first solid neurophysiological evidence of the Steadicam's extraordinary ability to generate a sensation of immersion in the spatiotemporal dimension of film, facilitating the spectator's identification with the camera's perspective through the incorporation of movements thanks to their embodied simulation stimulated by the activation of the mirror neurons. As the camera is strapped to the cameraman who is free to move as normal, its "point of view" assumes a corporeality that allows the spectator to turn his head, to walk toward something or someone. The fact that the camera is attached to and, indeed, actually becomes an integral part the cameraman's body, is of the utmost importance for this experiment.

These results were confirmed and extended by a second high-density EEG experiment in which we used short video clips showing an empty room filmed with a still camera, a zoom, and a Steadicam.[56] We found that event-related desynchronization of the beta components of the mu rhythm recorded by central electrodes was strongest for the clips filmed with the Steadicam. No equivalent modulation was found in the visual occipital areas, hence confirming the sensory–motor nature of the participants' neural responses to the video clips, in spite of the fact that, differently from the first experiment reported above, there were no acting agents to be seen. Moreover, in this experiment the video clips filmed with the Steadicam were experienced by all participants as best simulating a human approaching the scene. When watching a scene filmed with the Steadicam, we literally walk together with the camera's eye.

Vivian Sobchack grouped all camera movements under the label "bodily motions of the camera," whether the movements were to be attributed to a dolly or a crane, or to the cameraman himself; the notion that a camera moves through space, regardless of the technique used, has always given the idea that the spectator's motor engagement is intensified, which is why many filmmakers were not keen on the zoom. This is also due to the fact that, as a number of studies have shown, the average moviegoer is not always able to distinguish a

[56] Heimann et al. (2019).

dolly from a Steadicam.⁵⁷ Our two studies have shown for the first time that the motor cortical regions of the brain are perfectly capable of distinguishing not only between the zoom and actual movements of the camera but also between scenes shot with a dolly and a Steadicam. This greater resonance would seem to be attributable to the "biological flavor" of the movements of the Steadicam, anchored to the body of the cameraman.

The corporeality of the Steadicam has been exploited to the full in a number of famous sequences; for example, Martin Scorsese used the Steadicam in certain long tracking shots such as in the Copacabana Club scene in *Goodfellas* (1990) and in the counting room in *Casino* (1995). In both these scenes, the camera/spectator not only follows the characters, but also explores the profilmic space, turning its/his head and drawing/paying attention to certain details of the scene, or making bodily movements more or less related to those of the characters in that scene, alternating externally and internally generated moves in adherence to "Bob's Rules," of which we spoke earlier. The sense of immersion that this gives the spectator is made possible by the fluidity of the camera movements that produce an ecological approach to the scene without the jumpiness that is normally present in scenes shot with a hand-held camera. The sense of movement evoked by the Steadicam gives the impression that the spectator is moving independently within the scene, as evidenced by the stronger activation of the motor simulation resulting from the activation of the mirror neurons. The scenes shot with a Steadicam in *The Shining*, particularly those in the kitchen we described earlier, are an example of this.

While it is true that the Steadicam, as every other technique, can be used in a less engaging manner, playing a minor role in the embodied simulation processes, the results of our studies are in line with Vivian Sobchack's concepts, while they appear to be in contradiction with those expounded by Jean-Pierre Geuens, who held that use of the Steadicam renounces the force of the subjective stance, resulting in a totally disembodied version of viewing.⁵⁸

As stated previously, our study offers a feasible explanation in neurophysiological terms as to why many filmmakers favor the Steadicam over the zoom, which they find unnatural and forced. David Cronenberg remarked that "One tool I never use is the zoom lens because it doesn't correspond to my idea of filmmaking. The zoom is just an optical gadget; it's purely practical. And I will always prefer moving the camera, because I find that it physically projects you inside the film's space. And zooming doesn't achieve that. It keeps you

---

⁵⁷ See, for example, Geuens (1993–1994).
⁵⁸ Geuens (1993–1994, pp. 15–16).

outside."⁵⁹ Bernardo Bertolucci shared this opinion: "I hardly ever use a zoom. I don't know why, but I find that there's something fake about its movement."⁶⁰ As our studies show, the use of a Steadicam generates the strongest rapport between the spectators' motor involvement and camera movement.

The intentionality and subjectivity of a movie are largely based on the spectator's simulation of the camera movements, which would suggest that the immanent quality of cinematographic subjectivity lies mostly with the corporeal/bodily nature of the movie and its fruition. The results of these experiments provide a confirmation in neuroscientific terms for Dominique Chateau's statement on filmic subjectivity: "If film has something to do with subjectivity, it is to the extent that its moving form bears the imprint of a subjectivity."⁶¹ The relational nature of cinematographic style and the intersubjectivity of film can be profitably studied by focussing on the type of motor cognition stimulated in the spectator by the film techniques used.

The next chapter examines editing, another important aspect of style in film-making.

---

[59] Cronenberg, in Tirard (2002, p. 201).
[60] Cronenberg, in Tirard (2012, p. 109).
[61] Cited by Lageira (2011, p. 166).

4
Cut and Harmony

Calumet City

Calumet City, Illinois. A classical establishing shot of a bungalow sets the context for the scene (Fig. 4.1). *Cut.* A strange individual, the serial killer Jame "Buffalo Bill" Gump (Ted Levine), handling worms. *Cut.* A number of armed police officers in riot gear approach the bungalow and conceal themselves in the bushes nearby. *Cut.* Back to the man with the worms, who seems to have heard a noise and is looking at a point off-screen. *Cut.* The police officers in the bushes, filmed from a point close to the ground, an unusual angle. *Cut.* The man calls out to his dog, Precious, who is whining somewhere off-screen; visibly agitated, he starts moving through the building. *Cut.* More police officers arrive outside, on the other side of the bungalow. *Cut.* The killer enters the basement where he keeps his victims imprisoned and finds that his adored dog is being held by a young woman in a dry well in the floor. *Cut.* Jack Crawford (Scott Glenn), Agent-in-Charge of the Behavioural Science Unit of the FBI has arrived and is hiding behind a tree. *Cut.* The killer tries to persuade the woman to give him his dog. *Cut.* More police officers arrive. *Cut.* The killer continues to beg the woman to give him back his dog. *Cut.* A florist's van parks in front of the bungalow. *Cut.* The killer and the woman continue discussing the dog and pain. *Cut.* A police officer disguised as a florist opens the doors of the van and unloads a box, which seems to contain flowers. *Cut.* The killer and the woman continue their heated discussion, until the killer, furious, leaves the room. *Cut.* The outside of the bungalow once again as in the establishing shot, but this time the police officers and the van are present. *Cut.* Back inside the bungalow, tension is mounting. The woman yells that she will hurt the dog; the killer is in another room, loading a pistol. *Cut.* The camera is in the driver's cab of the van and shows a close-up of Crawford through the wound-down window and then a shot of the officer/florist ringing the door-bell of the bungalow. *Cut.* Detail of the internal mechanism of an old, rusty bell, ringing. *Cut.* The killer interrupts what he is doing and directs his gaze upward; he will need to go upstairs to see who is at the door. *Cut.* Outside, the officer in disguise rings the bell again. *Cut.* Another shot of the internal mechanism of the bell. *Cut.* The killer has made

The Empathic Screen: Cinema and Neuroscience. Vittorio Gallese and Michele Guerra, Oxford University Press (2020). © Oxford University Press.
DOI: 10.1093/oso/9780198793533.001.0001

Fig. 4.1 *The Silence of the Lambs* (Jonathan Demme, 1991).

CALUMET CITY 119

Fig. 4.1 Continued

himself as presentable as possible and appears in the kitchen, swearing under his breath he says he is coming. *Cut.* Crawford talking into a walkie-talkie, telling the officers they are going in; in the background other two officers take up their position with their rifles trained on the front door of the bungalow.

Cut. The killer reaches the door and opens it ... the spectator expects to see the officer disguised as the florist, but the person on the doorstep is FBI trainee, Clarice Starling (Jodie Foster). She is involved in the case, but the spectator knows she is alone and is not part of the team outside that up to that moment seemed to be preparing an assault on the killer's bungalow. Starling politely apologises for intruding, shows her identity badge—apparently not realizing that the man in front of her is the person she has been tracking for a long time—and tells the killer she is looking for the Lippman family. *Cut.* The police officers enter, battering down doors and breaking windows. *Cut.* On the doorstep, the killer tells Starling that the Lippmans don't live there anymore and goes to close the door, but she stops him and apologizing once again, repeats that she needs to talk to him. *Cut.* The officers are running through the bungalow, going from room to room. *Cut.* The killer asks Starling what the problem is and she replies that she is investigating the death of Fredrica Bimmel (one of the killer's victims). *Cut.* The officers haven't found any sign of life and communicate to Crawford that the house is empty; a slow camera move focuses on Crawford's face and shows the terror in his eyes. He is only able to murmur "Clarice." *Cut.* Starling explains about Bimmel, and the killer, after hesitating, says that maybe he can help her and invites her in.

What we have just described, omitting quite a few cuts from these two falsely parallel sequences of separate events, is one of the high-tension sequences in Jonathan Demme's masterpiece, *The Silence of the Lambs* (1991). In this case suspense is not generated by any special camera movements or overlapping gazes; it is created by a masterful mis-direction of what is known as continuity editing, unsettling the spectator's complete trust in this narrative technique.

Continuity editing is widely used in films, videomaking, and any product built with moving images. We come across narratives constructed with this technique every day; over time it has been seen to be the most effective way of ushering the spectator into narrative fiction, of immerging him in the plot. As George Wilson said, the purpose of these narrative forms is to create a sequence of "objective" shots, to make the intersubjective relations, the actions, and the situations in which the characters find themselves perfectly clear, and should this objectivity be lacking, to clearly mark the change in perspective (as is often the case, for example, with many point-of-view shots).[1]

As neuroscientists and psychologists have recently observed, the formal conventions on which this type of editing is based are compatible with the natural dynamics of attention and expectations of continuity in space, time, and

---

[1] Wilson (2006, p. 81).

action; the ways in which we are able to overlook the differences between film and reality provide an optimal sandbox for research, which could be extended to investigate how we use the same physical and cognitive processes employed in cinema in our daily perception of the real world.[2]

Continuity

Walter Murch, one of the greatest American movie editors, once said that after hundreds of millions of years during which life on Earth perceived the world as a continuous stream of linked images, all of a sudden, in the early days of the twentieth century, humans were confronted with something entirely new: Edited film.[3] Apart from some *ante litteram* editing (that can be seen in certain cycles of paintings), what Murch found quite remarkable is that in most cases the spectator does not see the cuts for what they are, perceiving the images in films as a continuous, transparent, narrative flow. He added that it would not have been surprising if the human brain had rejected editing, as it violates the continuity to which evolution and experience have accustomed us.

The issue of fragmentation and continuity has fascinated filmmakers and theoreticians from the very earliest days of cinema, as they tried to understand how it was possible that the spectator could perceive a movie as a continuous flow without noticing and objecting to the cuts. Joseph Anderson, who attempted to develop cognitive theories on cinema based on J.J. Gibson's ecological approach, observed how the veracity of the "filmic event" depends on a particular form of "cross-modal confirmation" that constructs the reality of that illusion from the very elements that should underlie its unnaturalness; the separation of the images and the almost constant use of music actually heighten the impression of reality instead of weakening it. According to Anderson, the power of film lies in the rules that cinema has constructed in a very brief space of time to conceal its true nature, resulting in a form of narrative construction that meets the criteria of our perceptive system, evoking a strong sense of unity and veracity.[4]

On the question of editing, Anderson agrees with J.J. Gibson's line of thought according to which "there is no need to perceive everything at once if everything can be perceived in succession,"[5] and proposes three fundamental rules

[2] See Smith et al. (2012) for an overview.
[3] Murch (1995, p. 19).
[4] Anderson (1996, p. 89).
[5] As is well known, Gibson dedicated a number of pages to "Cinema and Visual Awareness" at the end of his book, *The Ecological Approach to Visual Perception* (1979, pp. 292–302).

for cinematographic continuity; with some slight changes, these replicate the results of preceding theories and experiments.[6] The first is the rule that film manuals give as the basis for what Noël Burch called institutional mode of representation (IMR) and is constructed on connections, which is to say that the shots have to be connected by liaisons, i.e., the characters' body or eye movements, by diegetic or extra-diegetic sound effects, or by the position of the camera. If these connections are adhered to and jump cuts or evident changes in the camera's angle and perspective are avoided, the effect of continuity is guaranteed; the spectator will have the impression of a uniform flow of images and will generally be in a position to easily infer who is looking at what and to evaluate the characters' behavior and intentions without any particular effort. The second rule is based on physical orientation and is therefore related to the position of the character in the shot; although this may be illusory, it looks real to the spectator. Sequences constructed with the shot/reverse shot technique fall into this category. It is one of the methods most widely used to provide continuity in conversation, as is the over-the-shoulder shot. Anderson's third rule is based on large scale relationships between times, places, and events that can be far removed one from the other, but which can be brought together with parallel editing or flash back or flash forward.

Returning to the sequence examined at the beginning of this chapter, of course the spectator can infer what is likely to happen to the characters, to Starling in particular and even to the killer himself, from what he has already seen of the plot. From the moment in which one of those bungalows typical of certain American horror stories appears on the screen, and the filmmaker lingers a little longer than usual on the establishing shot, the spectator is drawn into a state of anticipation, and his expectations are of gradual increase of suspense. As in the majority of movies, the shot of the exterior of the building is followed by another showing the interior (remember *Notorious*?). The killer is inside. With the spectator now convinced that this bungalow is the killer's home (he infers this as a shot of the exterior of the building followed by a shot of the interior is consolidated practice in cinema to establish such an association), Demme and Craig McKay (his editor) decide to completely misdirect audiences with a parallel sequence of two unrelated events, using the rules of classical editing. The gaze connections are in place, as are those of direction, and the acoustics, which are extremely powerful in counterfeiting spatial juxtaposition and temporal continuity, are knitted together perfectly. The tension of the scene peaks when the killer goes to open the door, and when the spectator

---

[6] Anderson (1996, p. 92).

sees Starling standing in the doorway he is shocked into a new state of suspense. Starling is, in fact, an intruder in the editing, the spectator is no longer being shown a single scene processed according to the rules of cinematographic continuity but two scenes spliced in parallel editing and, what's more, that do not converge. They are two perfectly parallel narrative segments. The spectator experiences the same feelings of frustration as in the scene of the keys in *Notorious*, but this time it is the reaction to a narrative technique, extremely powerful in its deception.

Years before *The Silence of the Lambs* was filmed, before Anderson wrote his book, before the cognitive and ecological theories of cinema, the Russian filmmaker and theorist Lev Kulešov was conducting experiments to gain a better understanding of the secrets of the perception of continuity that were to have a significant impact not only on the practices, and the theoretical and aesthetic studies of cinema, but also on how the character's body is conceived and on the construction of filmic space. The titles of his experiments were all indicative of their content;[7] for example, in "Fabricated Person" he showed that by mounting together details of certain parts of the bodies of different women (eyes, mouth, back), he could deceive the spectator into thinking he was seeing just one person.[8] In his analogous experiment with landscape, "Fabricated Landscape," he demonstrated that places as far apart as Washington and Moscow can be perceived as being in perfect continuity and proximity if correlated and harmonized through editing characters' movements and actions. He stressed the point that it is the activities of the characters that connect and convey the impression of unity. He did something similar with the experiment he called "Dance" in which the dance is composed of shots and gestures filmed at different times and also from different angles, but which appear continuous due to the editing of the characters' movements. Finally, with the experiment entitled "Mozžuchin," he explained the effect that was to go down in history as the Kulešov Effect and is probably the reason why he is still remembered and his work still quoted today. He took a shot of a facial expression of an actor and juxtaposed it with other shots showing various subjects (a table laid with a plate and a bottle, a girl in a coffin, a woman lying languidly on a couch). Although the facial expression was exactly the same in each case, the spectators attributed it a different sentiment according to the object of the shot. While in the previous cases continuity was obtained through action, in the Kulešov effect it

---

[7] For further detail regarding Kulešov, see *L'Effet Kulešov/The Kulešov Effect*, particularly Kepley Jr. (1986).
[8] In so doing, he anticipated in film form Hans Bellmer's surrealistic chimeric dolls.

was obtained by inference; the first shot is attributed meaning a posteriori from the second, through cognitive processing[9].

Kulešov's experiments, and not just the paradigm that became famous as the Effect, show how the new spatiality and temporality created by editing rely on our perceptual and cognitive system and work perfectly if constructed to cater to the spectators' expectations and inferences. Of course, editing can also be much more than this and the new spatiality and temporality can have diverse impacts on the spectator, splitting sensations, producing metaphorical meanings that follow from the physical shock that the cut can provoke. In these cases, the primary objective is no longer transparent continuity but rather a tonal and rhythmic option, which as Sergei Ejzenštejn explained, starts from cinema's ability to involve the spectator on a motor and associative level and "intends to subject the public to a series of shocks that end up by producing the required overall emotional effect, applying the necessary pressure on the psyche."[10] The stimulus, which has to lead to an intellectual processing provoked by an intensification of the sensorial reaction to the rhythm of the film, is decisive for Ejzenštejn. In fact, he speaks of "the physiological sum total of the resonance of the shot" to evaluate the index of the impact that the editing will have on the spectator.[11]

Our ability to organize what we see space-wise and time-wise is linked to perception, to which must be added affection, cognition, and, above all, attention; that same attention Hugo Münsterberg identified as being the true driver of the effect filmic images have on the spectator and that Victor Freeburg explained with the example of how magicians deceive our senses; on the screen, it is not a question of the hand being quicker than the eye, but of the eye anticipating what the hand will do.[12] A form of control has to be added to perception, having to do with the actions or intentions that give the movie its backbone and with very precise narrative choices that will be honed by the editing techniques used. Ever since people started theorizing about films, they have been writing long lists of rules to define the ideal cut or to illustrate the ideology behind it.[13] Everyone, from Jean Epstein to Walter Murch, who has studied the rhythm of film, tends to insist on the necessity of restoring editing to its primary function as a narration technique to make the story flow, and while so doing remain aligned with the layout of the movie, the three-dimensional continuity of real

---

[9] The Kulešov Effect was recently investigated empirically by our group (see Calbi et al., 2017, 2019).
[10] Ejzenštejn (1986, p. 235).
[11] Ejzenštejn (1986, p. 56).
[12] Freeburg (1923, pp. 102–3).
[13] See, for example, Arnheim (1957, pp. 69–74).

space, and maintain focus on the primary objective of identifying and adhering to rhythmic variables that will imbue the sequences with emotion or mood (which was done to perfection in the sequence of *The Silence of the Lambs* discussed at the beginning of this chapter).

As Epstein wrote in 1923, there is an exterior aspect of the rhythm of filmic images that must not exceed the filmmaker's attention to "the psychological rhythm that translates into the rhythm of the life of the characters on the screen and of the screenplay."[14] In other words, the art of editing must not be an end in itself, but must always be in harmony with the requirements of the narrative it underpins. Epstein greatly admired the way American filmmakers moved "instinctively towards photogeny (the production of photographic images)," noting that as they worked from experience and empiricism, they didn't waste time creating a theory that made the concept of photogeny clearer, they were totally involved in creating dynamic images "that had not been carefully processed, but looked as if they had sprung spontaneously from nature," a fact that would give American cinema an "easy victory" with the majority of the audiences the world over.[15]

Although the actual term photogeny itself was not used, Epstein's concept, based on the harmonization of action and the effects of cinematographic continuity, was widely explored with a more didactic than cultural bias in the American literature on film as an art in the early 1910s; indeed, not without certain theoretic illuminations that Epstein and most European theorists were not aware of and, if the truth be told, have since largely been forgotten.[16]

Action, action, action!

There was a vast amount of literature on filmmaking published in the United States between 1912 and the first half of the 1920s, including many manuals for those who wanted to learn the art; institutes teaching cinema sprang up in many cities and there were even university courses on writing for cinema. In those years in America film was screenplay and the primary technique to be learned was *photoplay writing*; indeed, writing was studied more than the visual side of the craft and one of the most popular teaching journals, dedicated to imparting how to write for the cinema, was aptly named *The Photoplay*

---

[14] Epstein (2002, p. 43).
[15] Epstein (2002, p. 140).
[16] See Belton (1997), Polan (2007), Marcus (2007), Costa (2008, pp. 7-23), and Guerra (2013, pp. 7–40; 2014).

Author. Epes Winthrop Sargent, considered to be an authority on the subject, said that "the photoplay itself is one of the newest of the literary arts,"[17] but it was clear to all that just being knowledgeable about the "mechanics of motion"[18] wasn't enough to write stories suitable for the new medium; what was needed was a "photoplay mind."[19]

Writing a screenplay means imagining the written word transformed into images, incorporated into actions and gestures and the wider and more general movement of the movie. This feat of imagination is rendered even more difficult by the absence of a soundtrack, all the more so because the early American theoreticians/teachers of cinema were almost unanimously against the use of expository intertitles. The focus has to be on the theme of the action, how to represent the agentive continuity of the characters and the movie itself; to teach how to make a movie effectively, it is necessary to understand exactly what has to happen every time the filmmaker barks "Action!"

This command releases the real secret of the art of cinema, communicating directly through action. This assumes that the spectator is willing and able to process the actions he sees, to absorb them, adopt them, make them his, but above all to share them. Cinematographic action is the foundation for the intersubjectivity proposed by film narrative, embodied into the actors' gestures and behavior. However, it must also be coherent with film form, complying with action laws, but sometimes even overemphasizing the action in order to convey new meanings. Paul Auster expressed this concept with great clarity in his novel *The Book of Illusions* (2002), recounting what cinema was like in the beginning while telling the story of a silent movie actor who seems to have volatized into thin air.

> They had invented a syntax of the eye, a grammar of pure kinesis, and except for the costumes and the cars and the quaint furniture in the background, none of it could possibly grow old. It was thought translated into action, human will expressing itself through the human body, and therefore it was for all time.[20]

In those books written in America at the beginning of the 1900s and which were to play a crucial role in promoting the swift stabilization of the classical American style, the term "action" took on different meanings that were particularly relevant

---

[17] Sargent (1913, p. 8).
[18] Phillips (1914, p. 39).
[19] Taylor (1914).
[20] Auster (2002, p. 14).

for understanding the trends of the sense and the concept of action in cinema. It would even be correct to say the various meanings attributed to action over the years mark the evolution of cinema from the early days when a movie simply showed human beings or landscapes in movement, to what it is today, with action deployed in time and space to accomplish narrative purposes.

The books on cinema of this period frequently carry a glossary, so the various meanings attributed to action can be found and compared. In his widely quoted *The Technique of Photoplay* (1913), Sargent explains action as "(a) any gesture performed by the player. (b) the various actions of individual players whereby the narrative is advanced."[21] In other words, he is making a distinction between action in itself and per se, implemented and completed in the actor's performance (and, as such, immediately able to create resonance in the spectator) and continuous action expressing an open intentionality that facilitates the advancement of the narrative. In the first case, there is a concept of cinematographic action that takes form in the shot, transforming the actor into the element that embodies the action of the movie; in the second the concept of action takes form in the continuity between shots and so the constructed space-time of the movie becomes a virtual place in which the action is embodied. In both cases, action is the privileged means by which the spectator takes his place in the story, both through an embodied relationship with action in itself and per se and with its continuous declination, which is only apparently narratively disembodied. In an article published in 2014, we wrote that it is possible to identify at least three forms of embodiment in film that can be studied to understand the spectator's levels of resonance: The embodiment linked to the reactions to the bodies and objects that appear on the screen; the embodiment linked to the type of acting; and the embodiment related to the style chosen.[22] Those who studied action in the early days of cinema provide descriptions that already assume these forms of relation; it is also clear that even in those days it was understood that if a movie is to be a success with the public, spectators need to be able to decode the action.

Just a year after Sargent formulated his definition, Henry Albert Phillips, one of the American theoreticians who reflected at length on the concept of action, in his work *The Photodrama* (1914), defined it as the "specific act of a character; the scenario; combined elements that compose a drama."[23] According to Sargent, action is principally what the actor does, so the "body" in the scene

---

[21] Sargent (1913, p. 161).
[22] Gallese and Guerra (2014, p. 173).
[23] Phillips (1914, p. 213).

remains the focal point of the communication, while for Phillips it is necessary to be precise about the specificity of the action and accept that "action" can also be a series of combined elements that at a first glance may not appear to be action in the strict sense of the term.

It was becoming evident that the strength of the action alone, be it in an actor's single performance or triggered by the sight of a particular landscape, was not going to be enough to guarantee cinema lasting success; the notion was taking root that right from the scriptwriting phase a movie had to be plotted as a chain of actions that draws the story along and at the same time gives it continuity. In his book Sargent explains some tricks of the trade to the aspiring screenplay writer: Avoid psychological or poetic digressions that will only damage the construction of the movie, forget punctuation, just use a dash to give the idea of a chain of action, so that the scenario editor can visualize the movie without getting bored. He proposed something along these lines:

2 –Exterior Nell's home – Nell and Morton enter in car – leave car – exit into house

3 –Library – Nell and Morton enter – Morton angry – Nell defiant –Morton exits – Nell to desk – writes […] Nell calls maid – gives letter –maid exits – Nell cries[24]

What Sargent is, in fact, saying is that action is of prime importance for a movie, the unit of measure of filmic time, to the point that for the screenplay writer nothing must exist except a series of actions separated by a dash.[25] The true strength of cinema lies in representing "characters in action," a theme that was taken up by other authors and teachers of the period such as James Slevin, or certain supporters of "natural action" as a realistic base for films, such as Eustace Hale Ball, John Emerson, and Anita Loos, and which finally was developed by Freeburg.[26] According to Sargent, the rhythm of the movie (theories on this would be fine-tuned in both America and Europe just a few years later) was to be judged "roughly by the number of seconds it should take to play the action."[27]

---

[24] Sargent (1913, p. 41).
[25] James Slevin, in his *On Picture-Play Writing: A Handbook of Workmanship*, published a year before Sargent's better-known book, said that there are no hidebound rules to cinema except that there always have to be characters in action. (Slevin, 1912, p. 88).
[26] Slevin (1912); Ball (1915, p. 9 and following); Emerson and Loos (1921, pp. 10–11); Freeburg (1918, p. 83).
[27] Sargent (1913, p. 153).

Behind all this, however, lies in a nutshell one of the questions that has fascinated film theoreticians over the last decades and, precisely, the transition from bodily action as the motor and principle stimulus of the cinematographic plan, to the construction of rhetoric and iconography based on the semantization of the action vis-à-vis the plot and the style of the movie. The primal film action intended as a new search for a human form, already condemned to be submitted to the combined logic of realism and narration as Philippe Alain Michaud said,[28] is part of a phase in the history of cinema when the body was the unit of measure of the filmic space-time, on which (and not on an intentionality embodied in cinematographic action) a new spatiality and temporality was built. Jonathan Auerbach defined this relation between body and filmic space as *body shots*, in which the body and the actions it performs embody the dimension of the movie and lay the foundations of a new experience.[29]

However, to return to Sargent and 1913 when American cinema was maturing a standardization of style starting from the script, but also including the movement of the images, the actor was no longer the key element of the filmic experience, something else was needed, something that went beyond the chain of what André Gaudreault would have called "punctual" signifiers lined up in a row one beside the other.[30] Something had to be done to achieve a more sophisticated system of integrated narrative. Sargent attempted to match the corporeality of the action on the screen with the form of the new photoplay writing, seeking a relation between action as a new form of exhibition and as a form of narration. This was a delicate transition, done very pragmatically, which shows how cinema since the very beginning of the 20th century participated in this reconsideration of the potential of the actors' corporeality and motor plasticity (though this is rarely remembered today); in the theater this developed along the lines both of a refusal of dialogue and a prevalence of movement and gestures to indicate the space (the story narrated through action rather than words, Sargent would have said, or action clearer than words, according to William Morgan Hannon),[31] and of disjunction in the performance, to the point when the actor becomes the fundamental link of the new *mise-en-scène* and translates this function in stylistic terms.[32]

---

[28] Michaud (2000, p. 59).
[29] Auerbach (2007).
[30] Gaudreault (2008, p. 96).
[31] Hannon (1915, p. 30).
[32] In slightly different terms, the relation between action on the screen and movement was decisive for Ejzenštejn's theory (2009). See also De Gaetano for a less well-known analysis (2013).

Just one year before Sargent, Phillips was also writing about action, expanding the range of meanings of the term. Once again *action* meant the action of an actor, but this time with the addition of the meaning of screenplay (frequently known at the time as *Complete Action*) and *action* was also the *combined elements* that facilitated the narrative. He was referring to gestures and actions as compositional elements contributing to the construction of the movie, rather than in their primary significance. The fact that *action* was a synonym for screenplay is highly relevant; it was the most important part of the movie in those days and this development was to become decisive when, in a period of just four years, Freeburg extended the semantic value of the term by systematically substituting "writing" with "composition" to free the field from any possible doubts regarding the true nature of film, which is first and foremost a composition of forms imbued with their own life, dynamism, and power of expression.

When speaking of action, Phillips insisted on the great novelty of cinema compared to theater: What distinguishes film from theater (and literature) is the absence of dialogue and a different nature of thought, which in that phase and for technical reasons couldn't be expressed in words. With a statement that brings to mind the citation we borrowed from Paul Auster, Phillips writes that the task of the author (the screenwriter) can be summed up in four words, "clothing Thoughts in flesh."[33] Every single word in the screenplay has to be immediately translatable into action and not just through gestures, but through a line of action to be developed starting from the agentive forms of the actor's body as a "cell" of narrative continuity entrusted to the montage (the editing).[34] In Phillips' own words:

> The perfect photoplay leaves no doubts, offers no explanations, offers nothing it cannot finish. It is all action, action, ACTION! And by action we mean technically visualized interpretation of whatsoever nature that convincingly contributes to the perfect illusion of emotionally seeing a dramatic story.[35]

Of course, as this passage shows, the situation is rather more complex; the concept of action identifies totally with the concept of film, but Phillips talks of an action that can be of a different nature and has to facilitate illusion and

---

[33] Phillips (1914, p. 65).
[34] Phillips (1914, pp. 36–9, 84). See Levin et al. (2013) on the concept of continuity along the "action/basic agency/strong continuity" axis.
[35] Phillips (1914, p. 64).

emotion. In other words, the action must reinforce the intersubjective relationship between the spectator and the screen, eliminating the distances and boosting the emotion. All the same, as Phillips points out, it is not enough to leverage on "primal passion and primeval emotion," that is cinematographic attraction in and of itself, or what some of his colleagues called "punch" or "incident." The "action film" as described by Vachel Lindsay in *The Art of the Moving Picture* (1915) for instance, is the simplest type, and that most often seen in the early days of cinema. This genre became commonly known as "chase movie," a series of rocambolesque actions that burst on to the screen but lacked a real plot, petering out when there were no more stunts left. Action had to break free from this superficiality and insinuate itself into the dramaturgical structure of the movie, which, as we have seen, is action, action, action. The screenplay writer has to be able to represent "dramatic development visibly, in terms of action and symbols of emotion," keeping in mind that "the ability to represent this story in a form that may be readily interpreted depends on a practical assimilation and a working knowledge of dramatic construction and photoplay technique."[36]

According to Phillips, the key to the power of cinema lies in developing the movie along three basic modes, "psychological action, suggestive attitude, mimetic expression," rather more complex than the less defined categories (life, movement, and expression) that Sargent proposed earlier. Sargent discarded psychology on the same grounds that punctuation had been discarded—in his view, it was anti-cinematographic. Phillips retrieved it, dusted it down, and incorporated it into action, intention, and mimesis; now it is clear that mimesis cannot simply be a question of pure imitation, as it must regulate a narrative style whose continuity is spelt out by action, in an Aristotelian sense of the term that was very much in vogue in American cinema during this phase.[37] The importance of this psychological action is clearer if we think of the innovative value of the concept of "mental action" and "camera action" that Münsterberg initially developed in his article "Why We Go to the Movies" (1915) and took up again later in *The Photoplay: A Psychological Study* (1916).

Phillips explained how to obtain the effect of continuity with the "laws of natural movement and action," which on no condition must be violated if the story is to be comprehensible. What he was describing without knowing it

---

[36] Phillips (1914, p. 67).
[37] Slevin went so far as to mention Aristotle in his Acknowledgements in his *The Art of the Moving Picture* (1912, p. 5).

is one of the fundamental connections of classic découpage, the connecting movement and direction on which depends the "reflective power" that regulates the tempo and the accessibility of the scenes.[38] Hannon, who was greatly influenced by Phillips (his book was written a year after that of Phillips and shares the same title), returned to the theme of movement and action as key for the structuring of the plot, warning his readers not to fall into the error of the "chase movie," ALL ACTION instead of the action being modulated along a line of coherent dramatic development (now generally known as a "line of action").[39]

The line that leads from body movement to action and from action to continuity and then to narration was to a certain extent inherent in Sargent's dashes, a rejection of punctuation with the dashes representing the cuts (no other form of punctuation would have been appropriate for this function). This line was also present in Ball's attempt in 1913 to establish a length for the shots (this theme was subsequently developed in certain studies on style, cinematographic metrics, and the psychology of film reception, from Barry Salt to Yuri Tsivian to James Cutting),[40] and to underline the importance of close-up action,[41] only possible in film and which introduces a purely mental connection that would be impossible in nature and gives form to attention.[42]

It can be said that the concept of action is the basis of a reflection on the style of film that lead to the standardization of the classical American style, which is still the most widespread. In fact, the efforts of these early theoreticians/teachers practically coincided with the formative years of the classic style; they ceased when it was adequately fine-tuned and there was no need to work further on it. Continuity has been achieved by a process moving through body shots and attractions, a process that would be difficult to understand if action were not taken as the starting point. Indeed, as Roy Menarini wrote in an essay studying the forms of representations of the body in film, action "seems to coincide with cinema."[43]

The experiment we are about to share explores ways of relating to the space created by editing and the actions that take place within it.

---

[38] Phillips (1914, p. 102–4).

[39] Hannon (1915, pp. 38–9). However, it is known that the chase movie did play a role in the development of the causal line of action. See Gunning (1990, pp. 60–1).

[40] See Salt (1993). For Yuri Tsivian's project see www.cinemetrics.lv; Cutting et al. (2012). See also Pearlman (2009).

[41] Ball (1913, pp. 19–20).

[42] Münsterberg certainly knew Ball's book, as the copy in the Harvard College Library carries an ex-libris "From the Library of Hugo Münsterberg, Professor of Psychology."

[43] Menarini (2015, p. 15).

The mirror crack'd

Every movie is a concatenation of images, mostly linked by the continuity editing described in the previous section. As we saw at the beginning of the chapter, the editing makes it possible for us to lose ourselves in the movie's narrative fiction, blissfully unaware of the artificiality of what we are seeing; we are skilfully deceived by the techniques of classical editing. Editing permits the conjunction of shots and sequences that can differ greatly in their visual contents. An average Hollywoodian movie can contain up to a thousand different shots, an action film even double that. The sequence of a hand grasping a set of keys can be followed by that of a person entering a house. A hand seizing a revolver can be followed immediately by a body collapsing onto the floor, hit by a bullet from the self-same revolver from the previous sequence. The spatial–temporal and causal unity between the edited sequences is guaranteed, despite the fact that the perception of the flow is created from a succession of discontinuous images.

Given this, it is important to understand the characteristics of what our brain receives and processes from our daily observation of the world around us. The flow of images that reaches our eyes is continuously interrupted by our blinking; when we close our eyes involuntarily we are, in fact, lowering the curtain of our eyelids on the stage of the world outside. It is estimated that we blink 10–15 times a minute.[44] As to all extents and purposes we are blind when we blink and a blink lasts approximately 150 milliseconds, we lose from 1.5 to 2.2 seconds of vision each minute. And this is just part of the story! Every second our eyes make 2–5 rapid movements, known as saccadic eye movements, during each of which we are blind for approximately the same length of time as when we blink.[45]

From this point of view, our vision of reality and our vision of a movie share the same intermittency. Even in real life, what we think of as a continuous flow of images is, in fact, a succession of "snapshots" of the world, interspersed with black shots caused by blinking and saccadic eye movements. Approximately a third of the time we think we are looking at the world is, in fact, spent looking at a blank screen because our eyes are shut or busy with saccadic movements. In spite of this, our subjective experience of a continuous and coherent vision of the world is in no way compromised. Although it seems impossible that we should not be aware of these black shots, we seem to remap what appear as

---

[44] Burr (2005).
[45] Diamond et al. (2000).

disconnected images with the support of our ability to anticipate the existence, spatial localization, and contents of what we see on the basis of previous visual experiences.

Editing has put the nature of our vision to good use, exploiting these characteristics to enhance the impression of continuity that permits our immersion in the narrative of the movie, created by images, acoustics, and words. Numerous studies have shown that spectators viewing a movie, whether or not they are habitual cinema-goers and irrespective of their social and ethnical extraction, have no difficulty in understanding cinematographic narration produced with continuity editing.[46] It has been suggested that the apparent ease with which we are able to follow the plot in a movie derives from the fact that filmmakers apply a series of rules that regulate continuity editing.[47]

Various empirical studies have shown that during the viewing of a movie, blinking and saccadic eye movements tend to take place when the spectator's attention is at a low ebb, such as during the break between two events.[48] If the editing cut is made at this point, the spectator notices it less.[49]

The efficacy of the continuity editing technique depends greatly on the typology of the images prior to and following the cut. Successful masking depends on the extent to which the sequence that follows the cut is congruent with the expectations generated by what went before. One of the rules that governs continuous editing is closely related to the relationship between the predictive anticipation of what is about to appear on the screen and the continuous perception of the narrated events; known as the "180-degree rule," it concerns the position of the camera vis-à-vis the actors being filmed, establishing that the space in which the action is being filmed is divided in two by an imaginary line, the axis. The camera is placed in one half, while the other half is the so-called "pro-filmic space" in which the action is taking place (Fig. 4.2A).

The action develops as if on a stage and the camera is free to move between shots, the only condition being that it does not cross the imaginary line. Previous research studies have shown that editing adhering to the 180-degree rule does not generate a sense of discontinuity, while violating the rule does.[50] The explanation for this is that sequences containing camera movements adhering to the 180-degree rule are sufficiently similar to visual perception in real life, which is constantly being interrupted by blinking and saccadic eye

[46] Schwan and Ildirar (2010); Comuntzis (1987), Comuntzis and Page (1991).
[47] See Bordwell et al. (1985) and Cutting et al. (2012).
[48] Oh et al. (2012); Siegle et al. (2008); Nakano et al. (2009); Nakano and Kitazawa (2010).
[49] Smith and Henderson (2008).
[50] Zacks et al. (2009).

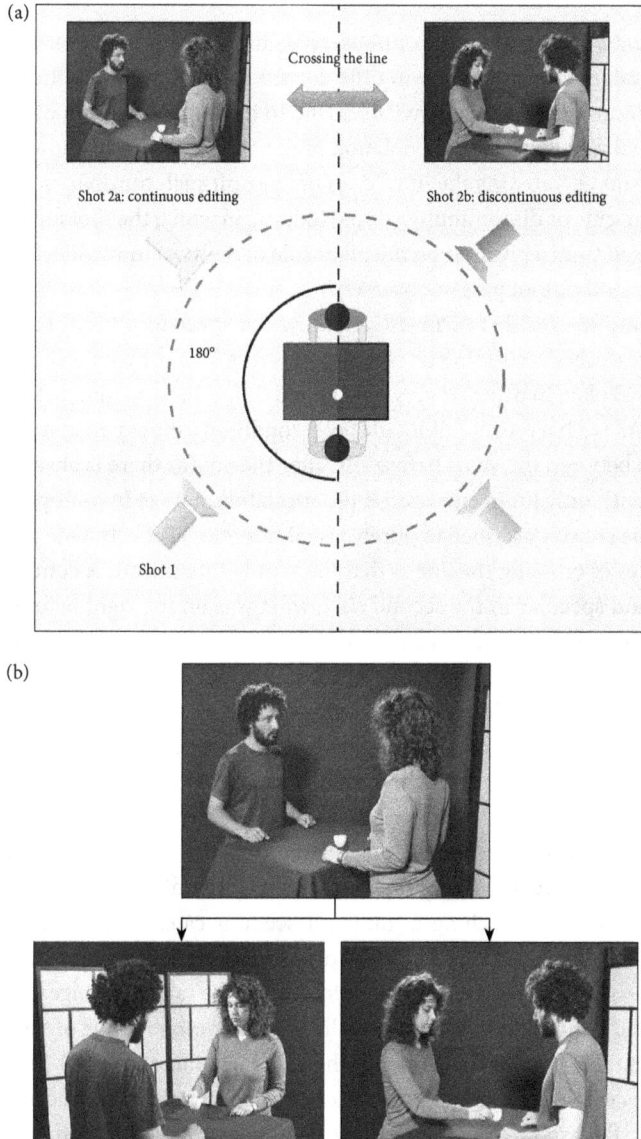

Fig. 4.2 (A) Schematic illustration of the 180-degree rule. From top to bottom: Crossing the line. Shot 2a: continuous editing; shot 2b: discontinuous editing (jump cut); shot 1. (B) Three photograms taken from the videoclip shown to the participants during the experiment. In continuous editing the first photogram is followed by the photogram on the left; in the jump cut condition the first photogram is followed by the photogram on the right. From left to right: 90° editing = continuous editing; 180° editing = discontinuous editing (jump cut).

movements, and so they pass unobserved. If the changes in position of the camera taking the shot following the cut are sufficiently small, they become "transparent" and therefore invisible. Due to its dramatic breach of our usual perceptual visual modalities, violation of the 180-degree rule, known in cinematic terms as "crossing the line," does not permit such transparency. As a result, jump cuts or discontinuity editing (editing in which the shot following the cut is taken from a position on the other side of the axis) immediately grab our attention and cannot pass unobserved.[51]

Recently we decided to investigate how the spectator's brain responds to sequences either respecting or violating this rule.[52] Our hypothesis, which we will now explain in depth, is very precise and differs in part from that described earlier. As we have seen, although both conditions show perceptual discontinuities between the shots before and after the cut (as there is always camera movement), only jump cuts arouse the spectator's notice. In our opinion, this cannot be entirely ascribed to purely visual-attentional effects. One of the consequences of crossing the line is that the shot before the cut is completely inverted and specular in the second shot; what was on the right before the cut, after the cut is on the left and vice versa. The spectator has the sensation that his perspective has been inverted. According to our hypothesis, therefore, the second shot, which violates the 180-degree rule, is not only profoundly disturbing from a visual point of view, but is also characterized by sensory–motor incongruence. This results in a temporary suspension of the embodied simulation through which the spectator becomes immersed in the scene, which means that his attention tends to focus more on the cut than on the content. Basically, what we are saying is that the perceptual dissonance the spectator experiences while watching a jump cut scene is caused by the violation of sensory–motor expectations generated by experience of corporeal interaction and the way things happen in the world. In fact, the radical change in perspective would require the spectator literally to jump from one side of the scene to the other. This implies that crossing the line, violation of the 180-degree rule, interferes with the functioning of the cerebral mechanisms that normally preside over the production of our actions and our observation of the actions of others.

---

[51] To minimize these effects, filmmakers often fall back on repeating the scene following the cut of the last part of the action shot before the cut. See Bordwell et al. (1985).

[52] This experiment was published in preliminary form in Katrin Heimann's doctorate thesis, conducted under Vittorio Gallese's supervision, and discussed successfully by Ms Heimann at the University of Parma in March 2015 (Heimann, 2015). It was subsequently published as a scientific paper in *Cognitive Science* (Heimann et al., 2017).

We used high-density electroencephalography (EEG) to verify our hypothesis empirically, recording the neural basis of perception of the two types of editing (continuous and discontinuous editing). We analyzed the event-related potentials (ERPs) and the event-related desynchronization (ERD) of the central mu rhythm and the posterior alpha rhythm during the observation of the first two seconds of the shot following the cut.

A group of 20 volunteers was shown short scenes involving a young man and a young woman. The scenes, each of which lasted five seconds, were produced editing two different shots (Fig. 4.2B). In the first shot, which lasted two seconds, the camera was positioned in front of the two actors who were facing each other across a table. In the second shot, which lasted three seconds, in 50 percent of the trials the conclusion of the scene was shot positioning the camera first on the near side of the axis (the imaginary line of the 180-degree rule), and on the far side of the axis for the remaining 50 percent. In the first scene the actors first look at each other, then one of the two lowers his gaze to an object on the table. Immediately after, the second actor also lowers his gaze to the object. At this point the shot stops. In the second shot the actor who had lowered his gaze to the object first, picks it up with his right hand and shows it to the other actor.

The variable in the experiment is the position of the camera from which the second shot of each scene was filmed. In 50 percent of the trials the camera was positioned at 90 degrees compared to the spatial position from which the first sequence was shot, thus adhering to the 180-degree rule; in the remaining 50 percent the camera was moved 180 degrees compared to the spatial position from which the first shot was filmed, therefore crossing the line. We used a total of 160 scenes; eight different scenes, each repeated 20 times in random succession, in which the actors alternated in grasping the object and showing it to the other.

The participants in the experiment were seated in front of a computer monitor on which the various scenes appeared. Each scene opened with a fixation cross that appeared in the center of the screen for a duration that varied between 500 and 1000 ms. This was followed by one of the eight different scenes, presented in random order. In 50 percent of the trials, the participants were asked to try to refrain from blinking while the scene was being shown, and to blink only when the screen went gray. After five seconds the fixation cross reappeared and another scene was shown. In the other 50 percent, a question regarding the scene was projected onto the screen when the scene ended and before the fixation cross appeared. The question was either related to the object (e.g., "Was the object on the table a coffee cup?") or to the actor (e.g., "Was the object shown to the actor or the actress?"). The participants responded to the question with a left click of the mouse, using their middle finger to indicate "yes," or a right-click of the mouse using the index finger of their right hand to indicate "no." The objective of the questions was to maintain the participants' high level of attention and to obtain data on which to compare the cerebral responses recorded during the participants' action execution and observation of the actions performed by the actors.

As previously mentioned, high-density EEG was used to record the participants' brain activity during the 160 trials; the results are illustrated in detail in the following paragraph.

ERPs are waveforms registered by EEG, evoked or correlated with a sensory input or a cognitive process. They are recorded by averaging differences of electric potential registered by electrodes placed on the scalp during the execution of trials of the same type (in this experiment, the sequences shot adhering to the 180-degree rule and when crossing the line), in order to improve the signal-to-noise ratio. Over the years various different types of ERP have been identified, differentiated by their polarity, which can be either negative or positive (respectively an upwards or downwards deflection of the electric potential), their latency (the moment in which the individual ERP triggers vis-à-vis when the stimulus starts to appear; in this experiment, this corresponded to the beginning of the second sequence), and, finally, on the basis of the spatial distribution and localization in the brain.

The results were particularly interesting. The first was the fact that the shots, when edited, evoked an early negative ERP in the electrodes positioned in the anterior section of the cerebral left hemisphere. This ERP, denoted N1, was recorded, on average, 140–190 milliseconds after the beginning of the second shot. Previous studies initially attributed this to the perception of syntactic irregularities of language; in these studies, N1 was frequently followed by a later positive potential recorded in the posterior electrodes attributed to a re-analysis of the linguistic structure.[53]

A more recent EEG study has investigated the syntactic violations of the structure of action, showing the participants actions taken from everyday life, but whose structure appeared different from usual due to an inversion of the temporal order in the presentation of two sequences normally adjacent within the same action.[54] This study has shown that the observation of syntactically

---

[53] See Hahne and Friederici (1999); Friederici et al. (1993); Friederici (2002); Grodzinsky and Santi (2008); and Steinhauer and Drury (2012).
[54] Maffongelli et al. (2015).

violated structures of action correlates with an early left anterior negative ERP, analogous to that recorded in our experiment. This potential appears to signal the breach of an expectation generated by the observation of the first scene preceding the cut. It is interesting to note that in the first phase, when the participants were watching the second shot, no significant differences were recorded between the two conditions (adhering to and violating the 180-degree rule). This suggests that whatever type of editing is used will evoke responses in the early phase when something is perceived to be out of line with what would have been expected on the basis of preceding perceptual experience.

The differences in watching scenes shot in adherence to the 180-degree rule and crossing the line only emerged later (Fig. 4.3). When our participants were watching the shots mounted with continuity editing, in which the 180-degree rule was adhered to, a late left anterior positive potential was recorded, starting 400–650 milliseconds after the start of the second shot. The amplitude of this potential, known as P4-6, was significantly lower when the participants were watching the shots with discontinuous editing. It is likely that the P4-6 potential we recorded correlated with an updating of the perceptual characteristics of the second shot, in an attempt to understand it better. This updating would be inhibited when watching shots edited with the jump cut technique.

When our participants watched the shots with discontinuous editing, which does not respect the 180-degree rule, a late positive potential known as P4-6 (400–650 ms after the start of the second shot) was recorded in the right parietal region. Earlier studies had recorded a late positive potential.[55] It has been suggested that this visual change in the observed stimuli could be associated with post-perception processes such as the conscious re-evaluation of what has been seen in order to decide what meaning should be attributed to it.[56] The higher amplitude of the P4-6 potential on the right parietal electrodes after viewing the shots mounted with discontinuous editing suggests that the participants were consciously reflecting on what they had just seen.

What is perhaps more relevant for our purposes is another interpretation given for P4-6 in the right parietal region; in fact, in earlier studies conducted to investigate control over the execution of the action and contemporaneous correction of errors, a late positive potential was also recorded by the electrodes in the central parietal region,[57] which increased in amplitude when an error in action was identified. Since this potential was noted in our experiment

---

[55] Niedeggen et al. (2001); Koivisto and Revonsuo (2003).
[56] Koivisto and Revonsuo (2003).
[57] Ruchsow et al. (2005); Falkenstein et al. (2000); Nieuwenhuis et al. (2001).

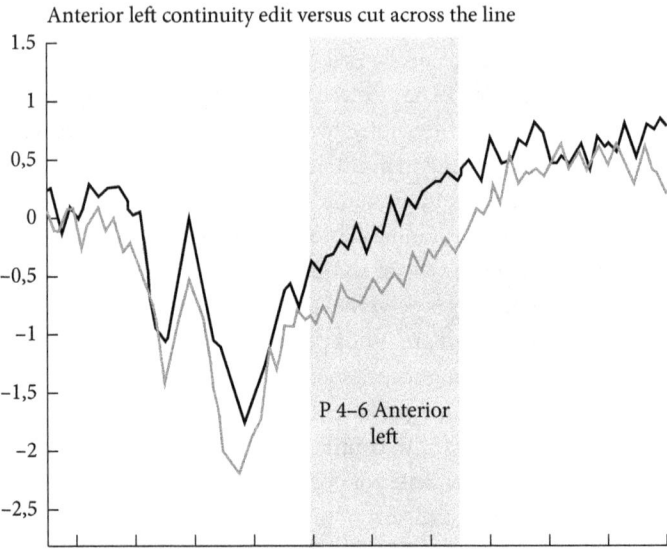
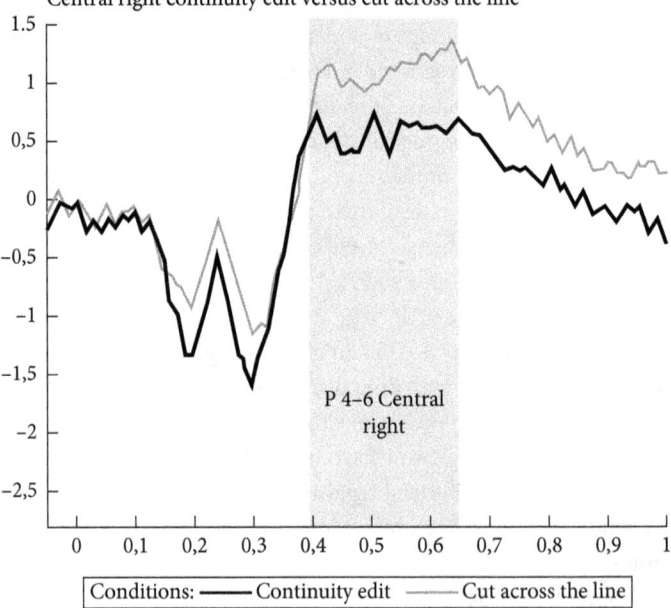

Fig. 4.3 Results of the event-related potentials (ERP) recorded from the left anterior and center right regions during observation of the two experimental conditions. The area shaded in gray shows the epoch in which the ERPs were significantly different in the two conditions ($p < 0.01$ for the anterior left and $p < 0.05$ for the center right) (Heimann, 2015; Heimann et al., 2017).

Left anterior region, continuous editing vs. crossing the line. Center right region, continuous editing vs. crossing the line. Conditions: continuous editing, crossing the line.

when the participants were observing the actions of others, it is possible that its functional significance is more general in nature, representing the correlate of the detection of an action that does not respect the observer's sensory–motor expectations, irrespective of whether the action was being actually carried out or just observed.

To summarize, the results obtained recording the participants' ERPs show an early N1 potential for all types of cut. In the past, this potential has been interpreted as the correlate of the violation of perceptual expectations, of ambiguity in what has been seen, or difficulty in understanding what has been seen. This confirms that when the spectator observes a visual discontinuity, such as that which occurs when there is any type of cut between two shots, his perception of the causal and spatial–temporal continuity of the scenes is altered. Whatever type of cut is used, it is perceived as a structural violation of the actions being observed. The absence of any significant differences between the two types of editing in these early components of the electrical cerebral response shows that in this initial phase of the observation of the scenes the brains of the participants had not picked up the difference in editing.

The main difference that emerged when we compared the two types of editing is the presence of the anterior left P4-6, which had a higher amplitude in the scenes with continuous editing, whereas when discontinuous editing was used, the posterior right P4-6 had a higher amplitude. In a functional magnetic resonance imaging (fMRI) study investigating editing of filmed scenes, Zacks and Magliano[58] proposed that the remapping processes involved, which in daily life are used to compensate small changes in visual input caused, for example, by body movements, are suppressed when the differences are greater, as when a jump cut is used. Here the differences are perceptually more salient and therefore would easily be spotted by the spectator.

On the one hand, our data are perfectly in line with this hypothesis, but on the other they are a step forward as they suggest that the perceptual incongruence generated by watching jump cut sequences is related to the individuation of a violation of the structure of the actions being watched. When watching an example of cinematographic editing, we use the same processes that we employ in our visual perception of the real world; however, depending on the type of editing used, we employ different late remapping processes. When we see scenes processed with discontinuous editing, these processes are inhibited to leave the field free for an analysis of the perceptual violation experienced, activating the circuits that preside over the execution and observation of actions.

---

[58] Zacks and Magliano (2011).

The second analysis in this experiment regarded event-related desynchronization (ERD), recorded while watching the same stimulus. As explained in the previous chapter, ERD is a quantitative change in the power of the frequency bands that characterize the rhythm of the EEG signal. The rhythms of these frequency bands are generated by the synchronization of a vast number of functionally correlated neurons at rest. When these neurons are activated by an external or internal stimulus, their synchronized activity stops and the rhythm recorded by the electrodes on the participants' scalps decreases in power.

The mirror mechanism for actions has been associated with the desynchronization of the central mu rhythm and therefore is recorded in correspondence with the motor cortices. Many researchers have confirmed that suppression of the central mu rhythm is triggered both by the execution of actions and by observation of actions being executed.[59] It has also been shown that the suppression of the central mu rhythm is greater in the hemisphere contralateral to the body part observed in action.[60] Another important rhythm is the alpha rhythm recorded by the occipital electrodes in correspondence with the cerebral visual areas. Desynchronization of this rhythm is generally interpreted as the correlate of the activation of the attentional processes.

Our experiment has shown a significant difference in the central mu rhythm when watching the two conditions (continuous editing and crossing the line) but did not reveal any difference in the occipital alpha wave (Fig. 4.4). While the participants were watching the shots edited with the continuous editing technique, a higher degree of activation of the mirror mechanism was recorded in the left hemisphere, which is the hemisphere that controls the movements of the right hand, and which in our case was the hand the actor used to grasp the object and show it to the other actor.

While the participants were watching the jump cut shots, the mirror mechanism was activated in both hemispheres. This lack of selectivity of the response vis-à-vis the hand the actor used to grasp the object is not surprising if we consider that in discontinuity editing, moving the camera's position produces an inversion of the spectator's perspective of the scene; what was on the right in the first shot, appears on the left in the second and vice versa. Our hypothesis is therefore that this causes transitory spatial disorientation, making it difficult for the spectator to identify the hand which actually performed the action as the right hand. As a result, the ERD recorded while the participants

---

[59] Derambure et al. (1993); Perry et al. (2010); Stancak and Pfurtscheller (1996); Toro et al. (1994); Pfurtscheller and Lopes da Silva (1999).
[60] Perry and Bentin (2009).

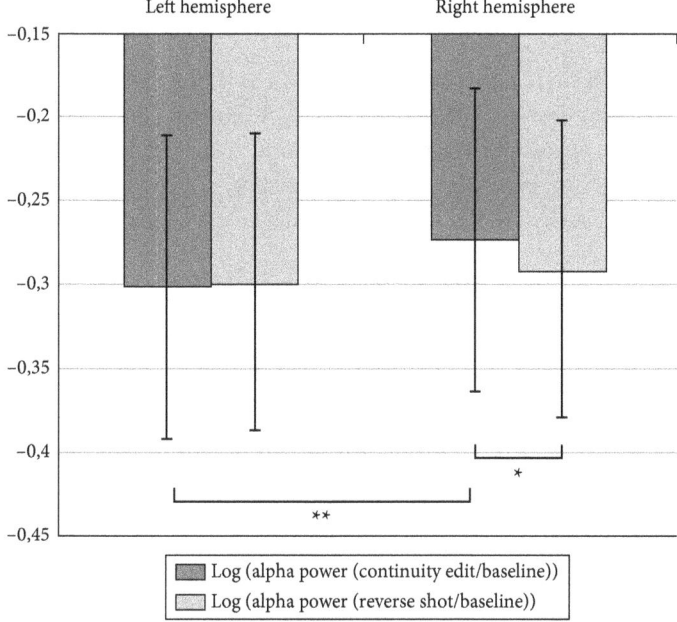

Fig. 4.4 Results of the desynchronization of the alpha rhythm in the two experimental conditions. The graph shows the significant differences recorded between the left and right hemispheres in the continuous editing condition ($p < 0.01$) and those between continuous editing and crossing the line in the right hemisphere ($p < 0.05$) (Heimann, 2015; Heimann et al., 2017).

were watching the jump cut sequences lacks the specificity of the side corresponding to the hand the actor used to grasp the object. No significant difference was found in the occipital alpha rhythm, which in our view is important as it would suggest that rather than depending primarily on visual attention, the perceptual analysis of sequences assembled with discontinuity editing depends on how far these sequences deviate from our visual perception in real life, or, in other words, on the extent to which the perspective from which we normally view the world is disoriented. The fact that ERD specificity was absent in the central mu rhythm, showing that our participants' simulation of the action they were watching was disturbed, could have caused the suppression of the remapping processes for the discontinuity editing (montage), as evidenced by the lesser amplitude of the left anterior P4-6 potential compared to the continuity editing condition. This, in turn, would favor the detection of a violation of the structure of the observed action, as evidenced by the greater amplitude of the right parietal P4-6 potential compared to the continuity editing condition.

To conclude, these results indicate that in filmmaking editing stimulates the perceptual competences and their underlying neural mechanisms that are normally used in visual interaction with the world around us. More specifically, editing engages an embodied visual perception that utilizes previous sensory–motor experiences from the real world to map, revise, and detect the discontinuities of the moving images in film.

The next chapter deals with other two dimensions that are crucial for the relationship between the spectator and the language of film: The "tactile" vision of what appears on the screen and close-ups.

5
Face and Hands

The empathy we experience with films shapes the way we understand them; and in turn, by focusing our critical attention on this empathy and the specific forms it takes, we gain a better understanding of the films themselves, whose meanings are conveyed not only through cognition and emotion but also through motion and physicality.

Jennifer Barker[1]

Touching in the mirror

An incandescent square of light exploding into a shot of a transparent strip of celluloid scrolling through an old projector, images of spools, a flash shot of an erection, more film, a single shot of figures moving in an old cartoon, a hand grasping the head of a lamb poised to perform the feral rite of sacrifice killing followed by a close-up of the intestines, hands nailed violently to the wood of a cross, a landscape that is the profile of the face of a corpse shot in close-up, a boy in various poses, sleeping, reading, who reaches out to touch the camera lens and so brings us, the spectators, closer to him. That same hand, after a cut that swings round 180-degrees, reaches out to touch the huge and initially out-of-focus faces of two women, who we then discover to be the characters of the movie (Fig. 5.1).

This is probably one of the most memorable incipits in the history of cinema. Few films are so strongly linked to the physical expressivity of the body and the materiality of objects as Ingmar Bergman's *Persona* (1966). He shows the human body metonymically through its essential components, the face, the hands. There are frequent close-ups of the faces of the two main characters, Nurse Alma, played by Bibi Andersson, and Elisabet Vogler, the stage actress, played by Liv Ullman; in one key sequence two halves of their faces are merged into one chimerical whole to suggest the deep sense of reciprocity of

---

[1] Barker (2009, p. 73).

The Empathic Screen: Cinema and Neuroscience. Vittorio Gallese and Michele Guerra, Oxford University Press (2020). © Oxford University Press.
DOI: 10.1093/oso/9780198793533.001.0001

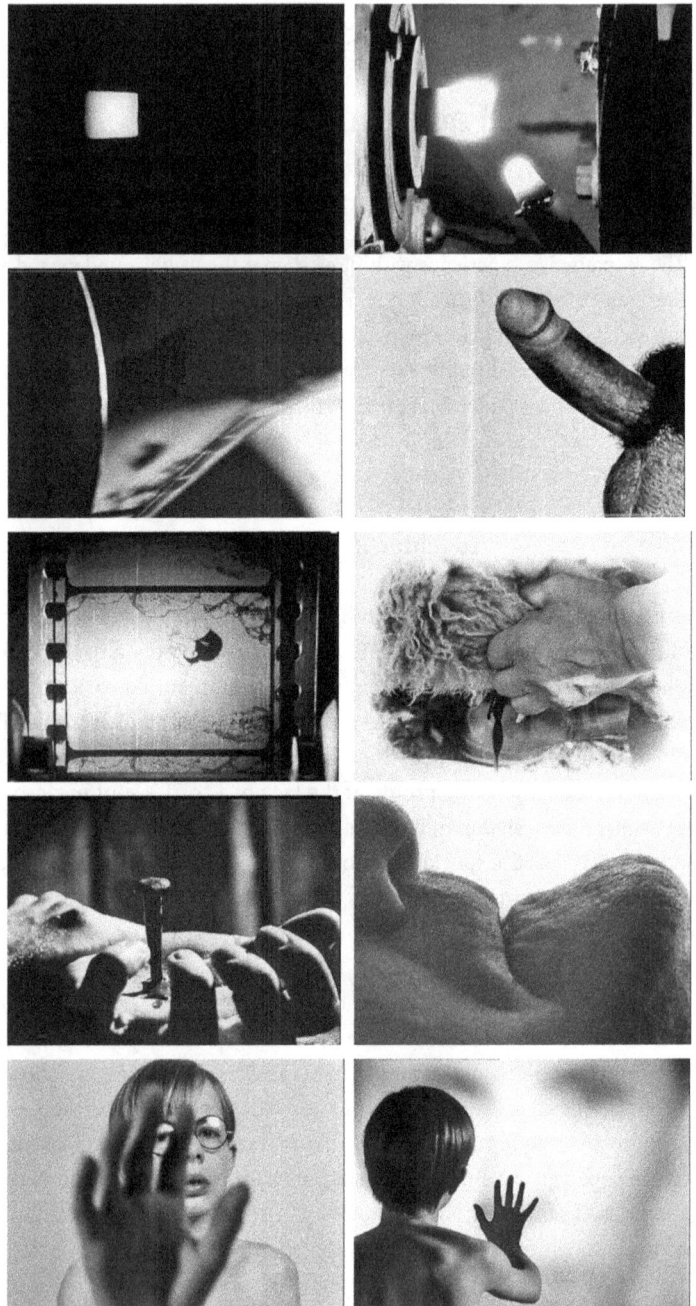

Fig. 5.1 *Persona* (Ingmar Bergman, 1966).

an intersubjective relationship that gradually, almost painfully, reveals their personalities.[2] The practically total absence of a soundtrack acts as counterpoint to Elisabet's silence and Alma's mounting loquacity, while Bergman displays the human body to emotionally stunning and dramatic effect, making surprisingly little use of cinema's classic means of expression: lighting, shots, and montage.

With *Persona*, Bergman takes the psychology of the characters and the ambiguous reception they receive from the spectator, setting everything in a meta-cinematographic meditation on the relationship between reality and representation, between public roles (actress, nurse, failed mother) and indefinable personal identities, between explicit narration and the undercurrent of pulsations and implicit memories that split coherency and modify the balance between dialogue and monologue. Initially Bergman intended to call it *Cinematograph*,[3] which testifies to the relationship between the movie's mystic, psychoanalytic, and religious undertones, and its exaggerated and explicit essence of "being film"; the original title effectively underlined the concept that the materiality of the relationships, starting from the faces and the hands, cannot be separated from the experience of the movie, which is presupposed by the sensorial and narrative workings of the camera. In a long essay published after *Persona*'s debut, Susan Sontag observed that the opening and closing sequences make it obvious that Bergman's intention was to "make the movie as an object,"[4] to make a statement of its concrete presence in our midst, its existence in space and time, but also its fragility, perishability.

In *Persona*, words are nothing and images are everything, the impotence of the former is offset by the brute force of the latter, to which Bergman entrusts the ambiguous truth of his message. The bodies of the characters and the physical world in which they move are caressed by light, shadow, and backlighting, so the tactile, almost prehensile, gaze of the camera with which we as spectators identify, is almost inescapable. This gaze is frequently reciprocated by Alma, Elisabet, and the young boy: Staring into the lens, pointing their own camera, or their hands, at the movie camera, they look at us and their gaze touches us to such an extent that it transforms into a hand reaching out to the camera lens.

Thomas Elsaesser and Malte Hagener analyzed this masterpiece of Swedish cinema with a particular focus on Bergman's expressive use of close-ups: A

---

[2] *Persona* incorporates Bergman's reflections on the human face; "the space of the virtual conjunction between the singularities" in the words of Deleuze, who describes the objective of Bergman's cinema as "the effacement of faces in nothingness" (Deleuze, 1983, pp. 127–8).
[3] Trasatti (1993, p. 85).
[4] Sontag (1969, p. 180).

female face projected "onto a translucent surface" and tentatively touched by the boy, pictures an archetypal relation enacted by cinema—that of serving as a mirror (mirroring as the moment in which a man confronts his real face).[5] For Gilles Deleuze, of all Bergman's movies *Persona* is the best suited to explain the director's typical philosophy of the face through close-ups, particularly the famous close-up of the merged faces of Alma and Elisabet:

> The close-up has merely pushed the face to those regions where the principle of individuation ceases to hold sway. They are not identical because they resemble each other but because they have lost individuation no less than socialisation and communication [...] [The close-up] absorbs two beings and absorbs them in the void. [...] The facial close-up is both the face and its effacement. [...] Bergman reached the extreme limit of affection-image [...].[6]

The effacement of the face is reflected in the decay of film and in the fragility of cinema. Cinema, to Bergman, is made for faces, for the affection/affectivity of close-ups. In an article he wrote in 1959 for *Cahiers du cinéma*, he said that "our work in film begins with the human face"; of course, it is possible to be carried away by the elegance of the editing, by the rhythmic staging of the objects and inanimate beings, by the beauty of nature, and so on, but "the possibility of drawing near to the human face is the primary originality and the distinctive quality of the cinema."[7]

This concept is deeply rooted in cinema, and has been since the early days of film. In the book Béla Balàzs wrote in 1924, he called the cinema an epic of sensations, primarily linked to facial expressions and to a physiognomy and mimicry that are decisive in theater but are much more so in film where the camera can come so close to the face that it is completely isolated. It follows that "close-ups are film's true terrain" and that "with the close-up the new territory of this new art opens up."[8] The concept itself of the close-up, of blowing up objects out-of-scale, isolating them from their normal environment until they are steeped in meaning, inevitably draws our attention to our perceptual relationship with the details of the faces or the grain of the objects that come under this magnifying glass. In 1916, well before Balàzs analyzed this concept in 1924, Hugo Münsterberg said that "the close-up has objectified our world of perception, our mental act of attention and by it has furnished art with a means

---

[5] Elsaesser and Hagener (2007, p. 57).
[6] Deleuze (1983, pp. 122–3).
[7] Bergman (1959, p. 150).
[8] Balàzs (1924, p. 175).

which far transcends the power of any theater stage."⁹ He saw a dual potential in the close-up, whether it be of a face, an object, or a limb: First, magnifying and getting closer to an image is an excellent way of facilitating forms of inference (forms that Münsterberg often refers to in his book as "mental processes") that originate in relevance and guarantee an increase in attention; second, the close-up acts as a sensory field of attraction and appeal that enhances the emotive and empathic force of the movie. Today we can say that there are two theories behind these potentialities of the close-up, one linked to cognitivism and traditional mind-reading,¹⁰ and the other to exploration of the precognitive and embodiment areas (or, as Torben Grodal would say, the approaches linked to the "Theory of Mind" and to the "Theories of Simulation").¹¹

Sergej Ejzenštejn himself had in mind a history of close-ups, reconsidering the function of *pars pro toto* in the history of art and re-contextualizing it in a study of what he called "sensitive thought." He was very attracted to the precognitive component of the close-up, which leaves an indelible mark on consciousness and memory ("my first conscious impression was a close-up").¹² As Antonio Somaini wrote, framing this study on close-ups within the great project of *Metod*, "according to Ejzenštejn the use of close-ups linked up to the reactivation of more efficacious structures of prelogical thought in the field of artistic creativity".¹³ In other words, the close-up intensifies the multimodality of our interaction with the movie¹⁴ and promotes a greater resonance not only with bodies and objects, but also with the *texture* of the movie¹⁵ that remains "beneath the representation" as a residual truth of the cinema, as is the case in Bergman's movie.¹⁶ From a different perspective, thinking of cinema as an excellent example of an art capable of eliciting forms of vitality through its techniques, Daniel Stern said that the intensity of the scene is proportional to the distance, and the levels of attention and arousal are regulated by how close or far removed the camera is positioned from the object it is filming: "Close-ups are of great impact because they breach the conventions that regulate bodily limits and the commonly accepted distances. They induce a form of arousal that prepares the body for any given form of activity

---

⁹ Münsterberg (1916, p. 88).
¹⁰ See Carl Plantinga on the field of cognitivism (1999, and more recently, 2015).
¹¹ Grodal (2009, pp. 143–4).
¹² Ejzenštejn (1995, p.21).
¹³ Somaini (2011, p. 242).
¹⁴ Marks (2002) adhered to this in several points of his study.
¹⁵ See Barker (2009, p. 74).
¹⁶ Definition of the "grain" of cinema offered by Guglielmo Pescatore, revisiting Epstein (Pescatore, 2001, p. 59).

(touching, kissing, beating, preparing to go to bed, etc.)"[17] David Bordwell reached similar conclusions in one of his major works on style in contemporary film.[18]

Throughout the 1900s and even in contemporary cinema, close-ups and detail maintain their original function of imparting visual shocks to the spectator; as Francesco Casetti pointed out, they help us see things better, but they also change the connotations of what we see and frequently show us impossible perspectives, extreme solutions, shots that are just too close for comfort. These are techniques that create within the plot "the convergence of the sensitive and the sensible [...] locus of a circularity between the two poles, both with a view to a re-functionalization of the sensible, and to a re-interrogation of the sensitive."[19] In a way this all leads back to Münsterberg, but it is very clear that a mere query of the narrative or of the emotional and affective salience is not sufficient in itself to explain the relevance and uniqueness of the close-ups, which—content and function apart—seem to have a much more crucial role to play, that of reinforcing the spectator's haptic and tactile resonance vis-à-vis the image on the screen. Although the literature on haptic vision (in which close-ups continue to be linked primarily with affection and identification) is constantly expanding, to date little has been written on this particular theme.

Taking the embodied simulation model as a starting point, our analysis lead to different conclusions. The body's symbolic translation into gestures and expressions betrays its intrinsic and dual theatrical nature; it literally stages subjectivity through a series of postures, sentiments, expressions, and modes of behavior. At the same time, by projecting into the external world, it renders subjectivity theatrical and transforms it into a stage on which corporeality is contemporaneously character and spectator, experienced and acknowledged. This relationship between the body and symbolic expression is the starting point for an investigation of the theme of artistic creativity expressed by cinema and its reception by the spectator.

This chapter is dedicated to an analysis of the tactile and haptic significance generated by close-ups not only of faces, but also of hands, bodies, and out-of-scale objects projected onto the screen with particular film techniques. Our hypothesis is that the close-up enhances the qualities of the object being filmed, such as anatomical details, weaves and wefts, the physical–material consistency of the image, to privilege the spectator's tactile and haptic resonance of the

---

[17] Stern (2010, p.81).
[18] Bordwell (2006).
[19] Casetti (2008, p. 135).

images through an enhanced evocation of embodied simulation. As we have seen in the preceding chapters, the immersive identification and the participation it generates vis-à-vis what we see on the screen, transit through a motor resonance with the movements, actions, and expressions of the actors. The mechanism itself does not require the image on the screen to be abnormally magnified, but it can be said that it activates when a full-figure shot of a character is presented; movements and actions are perceived and processed also when filmed with a medium long shot or a long shot.

In our view, the close-up enhances and focuses the spectator's gaze on the more material aspects of the object being filmed, whether it be a face, a hand, a landscape, a building, or an artefact. This is at least partially supported by recent neurophysiological discoveries demonstrating the multimodal nature of the somatosensory system. As explained in the first chapter, the cortical system that maps tactile sensations not only activates when we experience a form of contact, but also when we see another person having a similar experience.

The somatosensory system and multimodality

Neuroscientific research has shown that vision, touch, and action are indissolubly linked. When Alma places her hands on Elisabet's, which are concealing the photo of her son, our visual perception of their tactile experiences triggers the activation of our somatosensory and motor systems. In other words, we "see" the tactile experiences of others with our visual system, but also with our motor and tactile system.

Before going on to discuss this, a word of explanation regarding the organization and functional properties of the somatosensory cortical system that maps the tactile stimuli is in order (Fig. 5.2). The traditional concept holds that the primary somatosensory region only maps tactile and proprioceptive stimuli, the primary visual area only visual stimuli, and the primary acoustic area only auditory stimuli, but this has recently been challenged by a number of research studies according to which the primary sensory cortices (the cortical regions that process the most elementary characteristics of the sensory stimuli) are not exclusively unimodal, dedicated exclusively to a single sensorial modality.

With the support of contemporary cognitive neuroscience, it is possible to formulate a new perceptual model in which action, perception, and cognition are closely integrated, and in which multimodal integration, modeled on our corporeal experience of the world around us, provides input that our brain uses to map our existence.

152 FACE AND HANDS

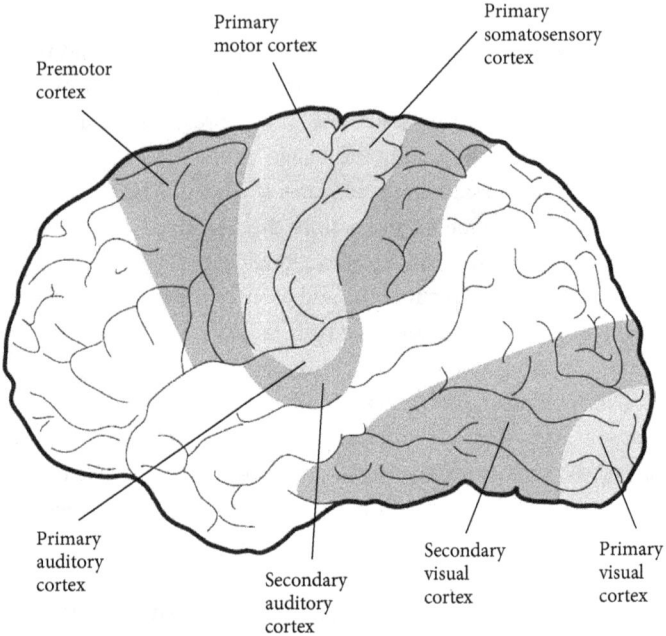

Fig. 5.2 Schematic representation of the human cerebral cortex, divided into the primary and secondary sensory areas, the motor and premotor areas.

Left to right, top to bottom: Premotor cortex, primary motor cortex, primary somatosensory cortex, primary visual cortex, secondary visual cortex, secondary auditory cortex, primary auditory cortex.

Applying the language of physiology to describe the stroke of a hand or a slap to the face, the caress and the slap become "tactile stimuli," mechanical events that take place on the periphery of our body where afferent or sensory neurons use receptor cells to transform the mechanical energy of the stimulus into action potentials (the electrical impulses produced by neural excitation). Numerous nerve fibers originate from the sensory receptors distributed throughout our body and carry these electric impulses into both the central nervous system and, in the case of stimuli coming from the head, face, and mouth, to the fifth cranial nerve, the trigeminal (Figs. 5.3–5.5). After passing through the brainstem and the thalamic nuclei specific for somatosensory sensibility, the fibers reach the neocortex, which in the case of humans is normally divided into diverse regions that are anatomically and functionally separate.

The primary somatosensory cortex (commonly known as SI) is situated behind the central sulcus that divides the frontal and parietal lobes; this is the

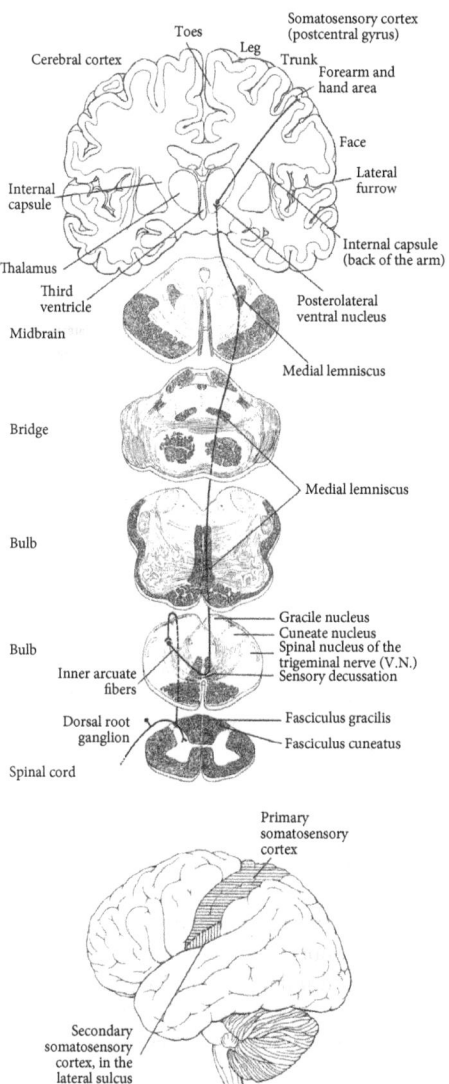

Fig. 5.3 General organization of the dorsal column–medial lemniscal system mediating tactile sensation and proprioception from the limbs. Three synapses are located along this pathway between the periphery and the cerebral cortex. The first is established by the centripetal branches of the fibers of the cells of the dorsal root ganglia with the neurons of the gracile and cuneate nuclei in the lower medulla. Axons of the neurons of these nuclei ascend the medial lemniscus and make contact with neurons of the ventral posterolateral nucleus of the thalamus. In turn, these thalamic nuclei project to somatosensory areas in the cerebral cortex. Below, lateral section of a cerebral hemisphere showing the location of the primary and secondary somatosensory cortices that receive projections from the posterior ventral nuclei.

Reproduced from Eric R. Kandel, James H. Schwartz, Thomas M. Jessell, Steven A. Siegelbaum, A. J. Hudspeth, ed., Principles of Neural Science, pp. 301–15, Figure 24.7 © 1985, Elsevier.

154   FACE AND HANDS

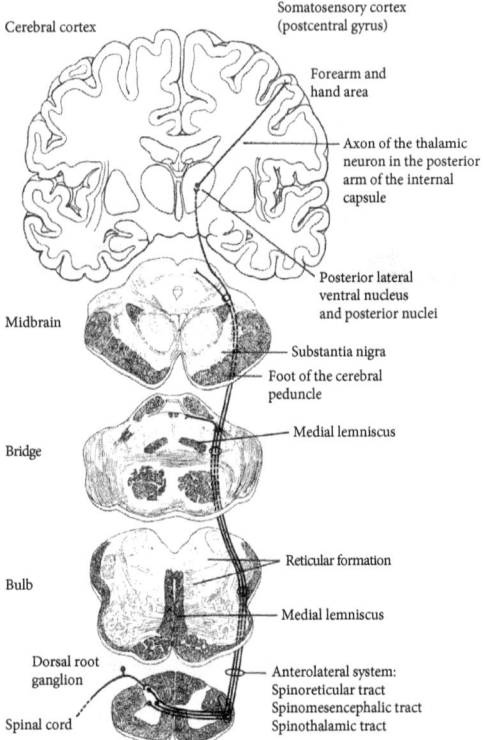

Fig. 5.4 General organization of the anterolateral system that mediates nociceptive and thermal sensibility. The three components of this system—the spinothalamic, spinoreticular, and spinotectal tracts—ascend the anterolateral cord of the white matter of the spinal cord. This system is organized somatotopically. The sensory fibers coming from progressively rostral areas gradually assume more ventral and medial positions.
Reproduced from Eric R. Kandel, James H. Schwartz, Thomas M. Jessell, Steven A. Siegelbaum, A. J. Hudspeth, ed., Principles of Neural Science, pp. 301-15, Figure 24.7 © 1985, Elsevier.

point of entrance into the neocortex for tactile stimuli. SI is composed of four distinct cytoarchitectonic areas: Brodmann areas 3a, 3b, 1, and 2. This cerebral region contains a somatotopical map of the tactile stimuli applied to the various regions of the contralateral side of the body.

"Somatotopy" literally means corporeal localization, the general principle of how the receptors are located vis-à-vis the corresponding areas in the cerebral cortex. Take touch, for example: The nerve fibers leading from the receptor cells distributed in areas close to the skin and the hypodermis travel in pairs to the brainstem, transiting various "stations" on the way and maintaining their

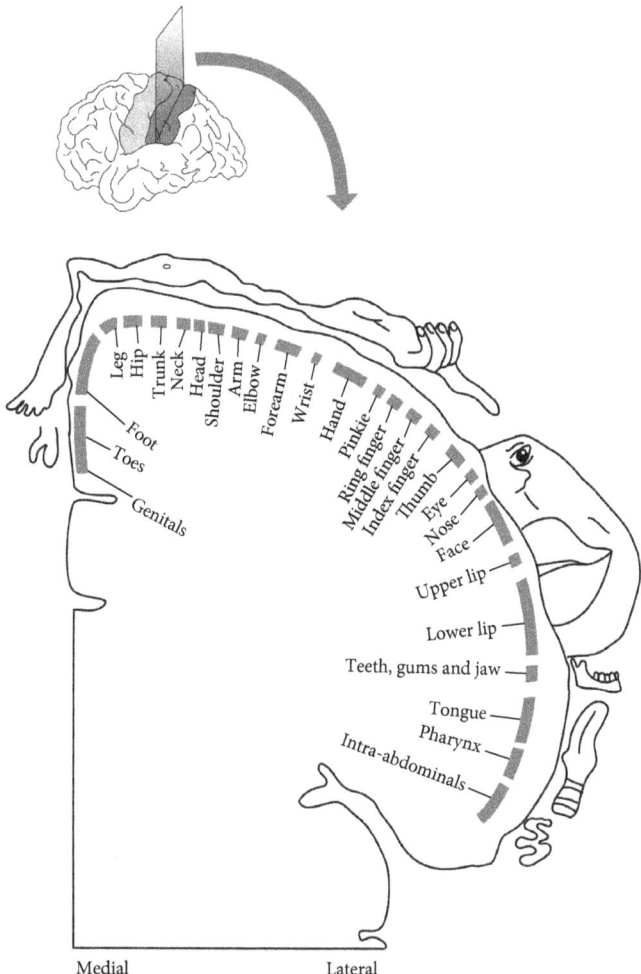

Fig. 5.5 Schematic illustration of the organization of the somatosensory homunculus in the primary somatosensory area SI

Bottom left to right: Medial ... genitals, toes, foot, leg, hip, trunk, neck, head, shoulder, arm, elbow, forearm, wrist, hand, pinkie, ring finger, middle finger, index finger, thumb, eye, nose, face, upper lip, lower lip, teeth, gums and jaw, tongue, pharynx, intra-abdominals... Lateral.

close relationship in each. The nerve fibers that arrive from the pinkie, for example, travel close to those coming from the ring finger, but both are far removed from those travelling from the thumb. This organization is maintained in the somatosensory cortices SI and SII, where there is a "neural map" of the body, mostly of its tactile sensations.

A similar organization is to be found in the primary motor cortex; however, here the neural map no longer plots sensations but the movements of the corresponding body parts. Indeed, the human body and the external world are plotted in the brain more than just once.

The secondary somatosensory area (SII) lies in the internal depths of the lateral sulcus and contains a less detailed bilateral map of the body. Both SI and SII receive somatosensory input from the thalamic nuclei of the somatosensory relay. Traditionally, it was thought that these two areas unimodally processed the somatosensory stimuli such as touch, proprioception, pain, hot, and cold, but this concept is no longer valid.

Over 30 years of research have shown that integration of the various sensory modalities is the rule and not the exception in humans and non-human primates. The visual areas also respond to tactile and auditory stimuli just as the somatosensory and acoustic areas respond also to visual stimuli. In addition, as we have seen, the motor areas contain motor neurons that also respond to sensory stimuli.

Consequently, it was necessary to substitute the single sensorial modality model subject to a process of association within the higher hierarchical cortical areas with another in which multimodal integration is a characteristic of the primary sensory areas.[20] Why do so many neurons in the human brain show multimodal integration properties? We suggest that the optimization of our ability to acquire knowledge of the world in which we live and move requires multimodality at neural level, as this is necessary to align the sensory systems of sight and touch with the motor system to guarantee that our behavior will be coherent in an equally coherent world that has meaning for us.[21]

The multimodality expressed by the motor and sensory neurons shows how neural integration is meaningfully guided by the characteristics of our bodily experience, rather than by an abstract and/or conventional computational logic.

In the past a clear distinction was made between unimodal modular structures for action and perception and the associative supramodal areas relative to the integration of the various sensory modalities. Empirical evidence suggests that sensory and motor systems are intrinsically multimodal and are directly interconnected, responding to and processing information associated to multifarious sensory modalities. This is certainly no coincidence; our bodily interactions with the external world, including those with other living beings,

---

[20] See Gallese and Ebisch (2013) for a discussion on this subject.
[21] Gallese and Sinigaglia (2011b).

are multimodal in nature. Consider what happens when we perform an action, such as grasping a cup of coffee; of course it involves the motor components such as movement, amplitude, direction, strength, and so on, but we must not forget the perceptual components such as the sight of our hand when we are in the act of grasping the cup, our perception of the cup itself, the proprioception relative to the position of our arm and hand at the instant in which we start the action, the position of the cup vis-à-vis our body, the sensation of heat emanating from the cup that we feel in our fingers as we grasp it, the sound of the spoon as we stir the sugar into the coffee, the aroma... ah, the aroma of that coffee!

The same logic applies to touch. The sensorial modality of touch, both when experienced subjectively and when we see it being experienced by others, is supported by dynamic processes of multimodal integration that include the activation of the somatic and visceral motor cortices.

We have seen that the motor and sensory cerebral cortices that directly drive our interactions with the external world with actions and perception also contribute to our conceptualization of what we see around us.[22]

In other words, these cerebral circuits express, from the neural point of view, representational content of what we see in the world around us generated by our body. This new model, which is substantiated by a significant body of empirical evidence, strongly challenges the traditional, but still dominant, classic cognitive science model according to which our conceptual knowledge of the world is supported by amodal representations from abstract and symbolic computations.[23]

As Edmund Husserl wrote, the body is "a thing inserted between the rest of the material world and the subjective sphere."[24] This body becomes *Leib*, by which is intended the "lived" body, containing its experiences of the world, which acts as a bridge between the subjective mental, physical, and objective dimensions.[25] Without this conception of the body it is not possible to account for the genesis of our experiences in the world or ourselves in that world.

Maurice Merleau-Ponty is extremely precise on the question of the Husserlian premises concerning the relationship between the body and experience. As he pointed out, "I perceive with my body" and "we perceive the world with our body. But by thus remaking contact with the body and with the world, we shall also rediscover ourselves, since, perceiving as we do with our body, the

---

[22] See Gallese and Lakoff (2005).
[23] Fodor (1975, 1983).
[24] Husserl (1952, p. 161).
[25] Husserl (1952, p. 161).

body is a natural self and, as it were, the subject of perception."[26] Consequently, the body overcomes the separation between physical and mental, "if we introduce the phenomenal body beside the objective one, if we make a knowing-body of it, and if, in short, we substitute for consciousness, as the subject of perception, existence, or being in the world through a body."[27]

In the following sections we will show how the sensory modality of touch is supported by dynamic processes of embodied simulation that include the activation of the motor, somatic, and visceral cortical system, both when our experience is direct and when we observe it in others.

The social perception of touch

Touch plays a particular role in our interactions with the external world. It is the first of the five senses to develop and is of paramount importance for babies and small children in discovering the animate and inanimate objects in their world and how they can interact with them.

It is especially important in social interaction; through somatosensory stimulation it facilitates non-verbal communication of intentions and affection in others. Unlike the other senses, the sense of touch is present throughout the body. Our bodily awareness of the external world, based as it is on both external (touch) and internal (the proprioception, originating from the articular, muscular, and tendinous receptors) perceptions, depends on the somatosensory system. As Husserl maintained, when we see something, we also see it as a tactile object, directly linked to the lived body (*Leib*), not exclusively due to its visibility.[28]

Recently much research has been done on the multifarious functions supported by the somatosensory system. Zhou and Fuster, for example, discovered that the SI neurons in monkeys also respond to visual stimuli if they have been previously associated with the touch of an object.[29]

The German philosopher Hans Jonas has suggested that the dynamic somatosensory feelings resulting from movement (such as fingers being pressed against resisting surfaces) are necessary to experience tactile qualities such as roughness or smoothness.[30] A subsequent study produced results that are coherent with this suggestion, discovering the presence of motor neurons linked

---

[26] Merleau-Ponty (1945, pp. 326–9).
[27] Merleau-Ponty (1945, p. 278).
[28] Husserl (1952).
[29] Zhou and Foster (2000).
[30] Jonas (1966).

to the prehension with the paw in the SII somatosensory area of the macaque.[31] As lesions in this area result in tactile agnosia[32] (the inability to recognize objects by tactile exploration),[33] these neurons might guarantee the conversion of the somatosensory input derived from the peripheral receptors that form a coherent image by moving one after the other when we explore an object with our hand.

When we reach into a pocket to extract our house keys, our hand only grasps the keys after having recognized them by a succession of exploratory movements. The motor system seems to supply the unifying criterion that ensures we can recognize the keys by tactile exploration without having actually seen them. The motor schema for grasping keys ensures that we attribute the meaning of "keys" to the tactile input generated by the movement of our fingers palpating the object in our pocket. This motor schema, which is different from that which guides our fingers in grasping the cell phone that occupies the same pocket as the keys, 'shows' our hand the object we want to grasp.

Moving into the field of social cognition, a series of functional magnetic resonance imaging (fMRI) studies have provided evidence that the sight of a person being touched activates part of those same cerebral areas, i.e., somatosensory areas SI, SII, and the premotor cortex (Fig. 5.6), that normally activate when we ourselves are being touched.[34]

Another fMRI study made a direct comparison between the observation of animate and inanimate touching, and stronger responses were detected in the SI area when the participants saw intentional and animate touching, such as a hand touching another hand as opposed to two inanimate objects accidentally touching, the branch of a plant blown by a ventilator, touching a chair (Fig. 5.7).[35] The intensity of neural activation in this area significantly correlated with the degree of intentionality of the observed touching. This finding suggests that in addition to the underpinning of the simulation of the tactile experiences of others, the SI area could also be involved in the simulation of the proprioceptive aspects related to our act of touching.

With respect to the affective aspects of social perception, the function of the somatosensory cortex has also been linked to empathic ability[36] or to the recognition of emotional expressions.[37] A number of studies have reported that

---

[31] Ishida et al. (2013).
[32] Caselli (1991); Reed and Caselli (1994, 1995).
[33] Valenza et al. (2001).
[34] Keysers et al. (2004); Blakemore et al. (2005); Ebisch et al. (2008, 2011); Schaefer et al. (2009); Meyer et al. (2011); Kuehn et al. (2013).
[35] Ebisch et al. (2008).
[36] Zaki et al. (2009); Schaefer et al. (2012).
[37] Adolphs et al. (2000); Pitcher et al. (2008).

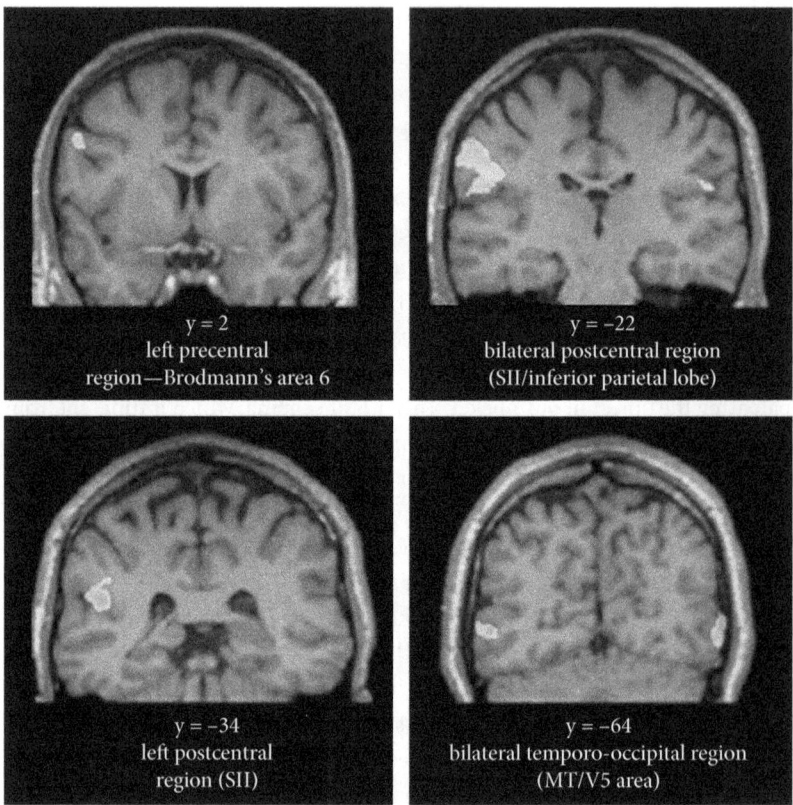

Fig. 5.6 Statistical maps showing the cerebral regions activated both by the subjective experience of a touch to the hand and by observation of someone else's hand being touched ($p<0.001$) (Ebisch et al., 2008).

the activation of both the somatosensory area SI[38] and the posterior region of area SII[39] is modulated by the affective strength or intensity of the touch in a social context: The sight of a caress or a slap on someone else's hand results in a greater activation of SII than the sight of a simple contact with no affective connotations (Fig. 5.8).

Ebisch et al. made another interesting discovery; the brain of a person watching a hand being touched differentiates between the condition in which the contact is experienced subjectively and when the contact with

---

[38] Bufalari et al. (2007); Bolognini et al. (2013).
[39] Ebisch et al. (2011).

THE SOCIAL PERCEPTION OF TOUCH 161

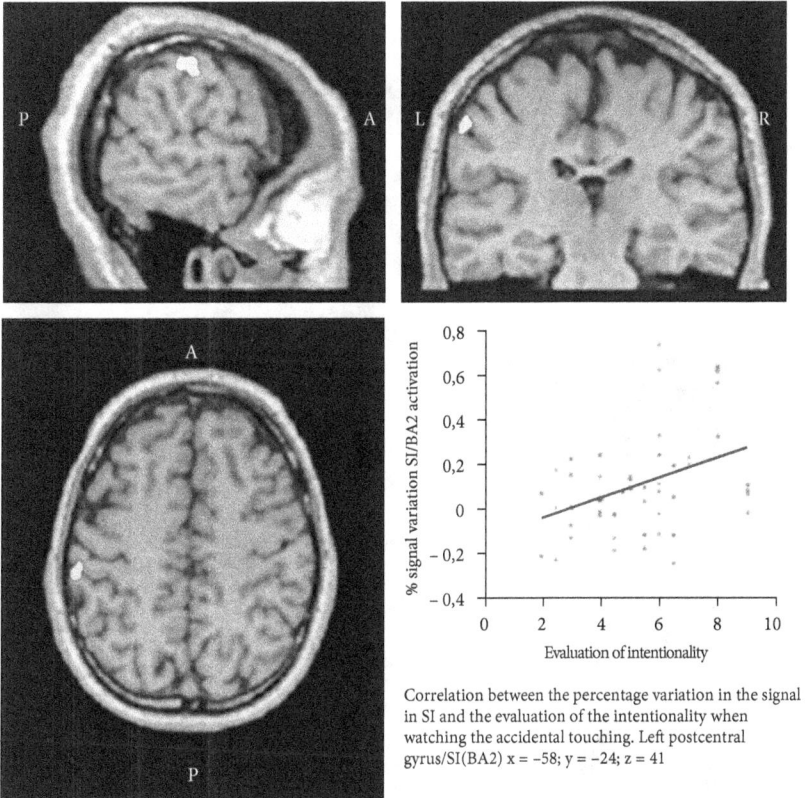

Correlation between the percentage variation in the signal in SI and the evaluation of the intentionality when watching the accidental touching. Left postcentral gyrus/SI(BA2) x = –58; y = –24; z = 41

Fig. 5.7 Statistical maps showing the increase in the activation ($p<0.001$) in SI (Brodmann's area 2 (BA2)) while watching intentional, as opposed to accidental, tactile stimulations (Ebisch et al., 2008).
P, posterior; A, anterior; L, left; R, right.

the hand of another person occurs on the screen. In the first case, the posterior insula activates, whilst in the second it deactivates. When we see other people being touched, we simulate this experience activating parts of those cerebral areas that we normally activate when we ourselves are touched in the same places. Other cerebral areas such as the posterior insula, however, behave in the opposite manner so that the sensation that has been observed is correctly attributed to the other person and not to ourselves.

Finally, other studies have shown that the somatosensory area SII also activates when the touching concerns inanimate objects, when one object touches

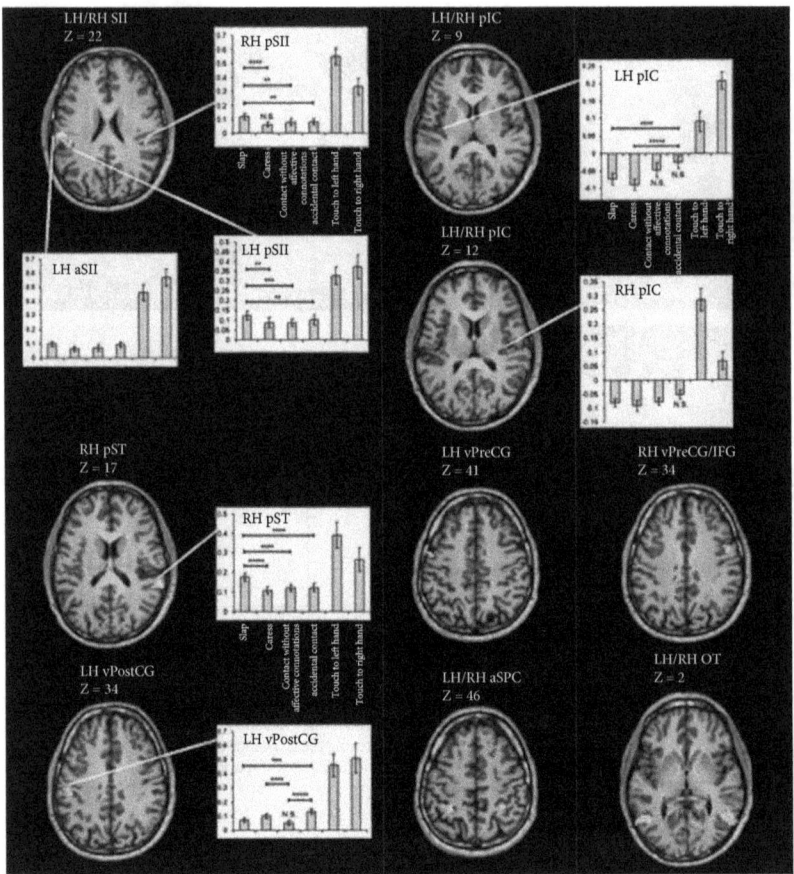

Fig. 5.8 Statistical maps showing the overlapping of the increase in activation during a subjective tactile experience and while watching it happening to another person (light gray), and the decrease in activation while watching it happening to another person (dark gray). The graphs show the percentage variation in the signal for the various tactile observation conditions (slap, caress, contact without affective connotations, accidental contact) and for the subjectively experienced touch to the right and left hands (**$p<0.05$; ***$p<0.05$; ****$p<0.005$) (Ebisch et al., 2011).

LH, left hemisphere; RH, right hemisphere; pSII, posterior secondary somatosensory cortex; pIC, posterior insular cortex; pST, posterior superior temporal gyrus/sulcus; vPreCG, ventral precentral gyrus; vPreCG/IFG, ventral precentral gyrus/inferior frontal gyrus; vPostCG, ventral postcentral gyrus; aSPC, anterior superior parietal cortex; OT, occipito-temporal junction.

another.[40] Ebisch et al. showed that an automatic tendency to activate the brain areas involved in the processing of our own experience of touch applies to the observation of any type of touch.[41] Activation of SII occurred independently of whether the touching observed was intentional or accidental, and also independently of the typology of the object involved, whether animate or inanimate. This would suggest that the principles of embodied simulation apply also to the understanding of abstract events, as well as to actions and corporeal sensations. It is not what is touched that is important, but the observation of the contact between two objects, whether they be a body part or an inanimate object. Even the conceptualization of an abstract notion like "contact" could, at least in part, be supported by embodied simulation processes.[42]

This hypothesis found corroboration in a recent fMRI study which demonstrated that the SI somatosensory cortex activated while reading linguistic metaphors concerned with texture, suggesting that comprehension of these metaphors could be perceptually based on embodied simulation in the sensory systems. It is conceivable that abstraction is a prelinguistic outcome of a multimodal integration mechanism.[43] It is interesting to note that as far back as the eighteenth century the Italian philosopher Gianbattista Vico wrote that "...n'tutte le lingue la maggior parte dell'espressioni d'intorno a cose inanimate son fatte con trasporti del corpo umano e delle sue parti e degli umani sensi e dell'umane passioni" (it is noteworthy that in all languages the greater part of expressions relating to inanimate things are formed by *transpositions* from the human body and its parts and from the human senses and passions).[44]

In conclusion, although it is evident that the somatosensory system is endowed with tactile properties, the data summarized here suggest that its functions extend well beyond its classic role in the perception of somatic sensations. On the question of the social domain, it is highly probable that embodied simulation gives us the possibility to map the bodily experiences of other people by reusing our own somatosensory representations in bodily format. By exploiting the same neural circuits as those recruited for subjective bodily experiences, it is possible to establish a direct intersubjective link between ourselves and others, and so we are able to have an experiential understanding of their tactile experiences.[45] We suggest that this mechanism plays an important role in determining the tactile quality of moving images on the cinema screen.

[40] Keysers et al. (2004).
[41] Ebisch et al. (2008).
[42] Keysers et al. (2004); Gallese (2005).
[43] See Gallese and Lakoff (2005); Gallese (2008).
[44] Vico (1725/1744, Bergin Fisch translation, Chapter 2, p. 116, section 405).
[45] Gallese (2003).

Feeling the film

In a book she wrote almost 20 years ago, Laura Marks proposed studying film as a sensitive element that induces embodied sensations and memories in the spectator, capable of reshaping cultural trends in the creation and reception of moving images. The book is entitled *The Skin of Film*, a metaphor that, for Marks, was to be taken literally: Films and videos can be considered as sensitive as the skin, with all the resulting consequences on the level of physiology and embodied experience. She speaks of "tactile images," referring to the fact that sight is directly related to touch and so exposes the spectator to a form of *haptic visuality*, almost as if he touches the film with his eyes.[46] In a study conducted some years after the publication of Marks' book, Jennifer Barker hypothesized that the haptic visuality to which Marks referred and which is also to be found in Vivian Sobchack's phenomenological reflections, responds above all to our innate need of contact, our tendency to touch others and act on the base of that stimulus; according to Barker, cinematographic spatiality is not exclusively visual, it is also tactile.[47] Touch is the sense that welds the spectator's experience even more firmly to the "flesh" of the images and evokes sensations and memories that can be either congruent with the plot or perturbing, capable of implementing the multimodality of the filmic experience. Touch adds that extra something to the reality to which Christian Metz was referring when he wrote that movement is insubstantial and "we see it, but it can never be touched," where the tactile sense is "the supreme arbiter of reality," that assumes the role of a criteria of materiality, against the insubstantiality and visuality of movement.[48] Research in the field of phenomenology has had a strong impact on the concept of tactile experience and the so-called virtual tactility; the neuroscientific evidence from this research is fundamental for a new development of Metz's position and a re-discussion of *haptic visuality*. As we have already said, in our view close-ups and details provide the maximum stimulation for tactile resonance mediated by film; we are stimulated more by the grain of the image than by what we actually see and the plot itself. We resonate at the sight of magnified or fragmented bodies or the skin of characters in close-up, and this brings us into carnal contact with the texture of the image. In this sense, Bergman's idea of linking the significance of *Persona* to the nature of the filmic image is somehow paradigmatic.

---

[46] Marks (2000, pp. XI–XII).
[47] Barker (2009, pp. 37–8).
[48] Metz (1968, pp. 37–8).

At the end of the 1970s Jean-Luc Godard held a series of lectures at the Concordia University of Montréal entitled *Introduction à une veritable histoire du cinéma*, which were then brought together in a book to constitute a viaticum for his major project *Histoire(s) du cinéma*.[49] In the mornings Godard scheduled projections of various sequences from masterpieces of the history of cinema related for one reason or another to his own concept of cinema and his own work, whilst during the afternoons he guided a discussion of the morning's viewing. Bergman's *Persona* was in one of his personal programs; playing with cinema in his own way, Godard said he had made a mistake, that he thought another of Bergman's films, *The Silence* (*Tystnaden*, 1963), was titled *Persona*, saying "I don't know the history of cinema very much anymore";[50] this mistake, whether genuine or not, helps our case, as he projected *Persona* immediately before *A Married Woman* (*Une femme mariée*, 1964), when in reality the latter preceded the former by two years. Godard confessed that he was inspired by Bergman's use of images and spoke of *A Married Woman* in these terms to *Cahiers du cinéma*:

> Woe unto me, then, since I have just made *A Married Woman*, a film where subjects are seen as objects, where pursuits by taxi alternate with ethnological interviews, where the spectacle of life finally mingles with its analysis: a film, in short, where the cinema plays happily, delighted to be only what it is.[51]

This makes it quite clear that the story, 24 hours in the life of Charlotte (Macha Meril), who divides her love and body between Robert (Bernard Noël), her lover and Pierre (Philippe Leroy), her husband, is basically a collection of "fragments" of life, which by deconstructing and then recomposing forms of desire (for sex, money, beautiful women), are seen for what they really are: Fragments of film (in fact the subtitle of the movie is "*Suite de fragments d'un film tourné en 1964*"). The characters are treated as objects that the camera films at will, sliding over surfaces, forcing the spectator to what can be called an epidermic face-off. For Godard, representation mingles with analysis and the sensitive function of the image oscillates between these two poles; however, this oscillation must be accepted for what it is, a manifestation of the freedom of cinema; it is freedom of expression, of use of the texture of images free of narrative constrictions and over and beyond the contemplative option.

---

[49] Godard (1980).
[50] Godard (1980, p. 119).
[51] Now in Godard (1985, p. 65).

166   FACE AND HANDS

The famous opening scenes of the movie, that, according to Alberto Farassino, offer "the most delicate and enchanting love scenes in contemporary cinema,"[52] constitute an excellent example of the freedom of cinema, at the same time offering a reference point for the forms of tactile resonance we are discussing. This also provides an opportunity to examine a typical tendency of cinema (and not only)[53] in the 1960s to create a contact with bodies, faces, with the surfaces of objects, almost as if this attention to the skin and to surfaces would sharpen the tactility of the relative images, while communicating their essentially cinematographic nature. It is no surely coincidence, therefore, that in Godard's lectures *A Married Woman* is juxtaposed with Bergman's *Persona* that debuted slightly later, nor that within the more complex relationship between Godard and Alain Resnais a point of contact can be found between this incipit and that of *Hiroshima mon amour* (1959).[54]

The first shot of the movie is a perfect blank, as if it were intended to duplicate a cinema screen (Fig. 5.9).

Then a woman's hand appears from below, a wedding ring on her finger, gently caressing the white surface, sliding up sinuously until her whole arm, naked, is in the shot. A man's hand appears top right, after which his forearm gradually enters the shot and the hand grasps the woman's wrist while we hear the couple, Charlotte and Robert, talking of their love affair off-screen. The next shot shows the woman's back; we can't see her face, but we can admire her page-boy haircut. The man's hand appears suddenly on the woman's shoulder, caressing her skin, passing under her armpit and coming to rest on her side. The man wants to know why the woman has a mark on her skin, and the woman touches her shoulder and back with her hands. The shot fades out through black, leading to the third shot. Now we see Charlotte, nude and in profile, sitting on the bed with her knees touching Robert's; Robert is sitting in the same position but we can only see his hairy knees. Then once again Robert's hands enter the shot, brushing Charlotte's face and legs, while she in turn caresses him, speaking of their love. The fourth shot: a close-up of Charlotte, we see her face and shoulders, pale against a white and neutral background. Suddenly Robert's hands appear again, sliding up from her breasts (which remain off-screen) to encircle her neck, and then start caressing her shoulders. He tells her she should be like the women in Italian films, who don't

---

[52] Farassino (1974, p. 82).
[53] Referring specifically to *A Married Woman*, Luigi Allegri suggested considering the ways in which human relationships, bodies, and objects are represented in the light of the new dada and pop cultures that burst onto the scene at the Venice Biennale of 1964 (Allegri, 1976, p. 111).
[54] Liandrat-Guigues and Leutrat (1994, p. 56).

shave their armpits. She retorts that she prefers American films, which are more entertaining. Yes, Robert replies, but much less exciting. The shot fades out through black bringing us to the fifth shot, which shows Charlotte's nude leg from thigh to foot; Robert's hand erupts into the shot, touches Charlotte's thigh, passes to the knee then lifts the sheet to reveal her other leg and his hairy leg. Charlotte tries to pull the sheet up to cover herself, and Robert asks her why she doesn't want to be looked at. Sixth shot: A close-up of Charlotte's flat, white stomach against a white and neutral background; the darkness of her belly-button is the focus of the scene. Robert's hands appear from the bottom of the shot, caressing her skin and starting to massage her stomach.

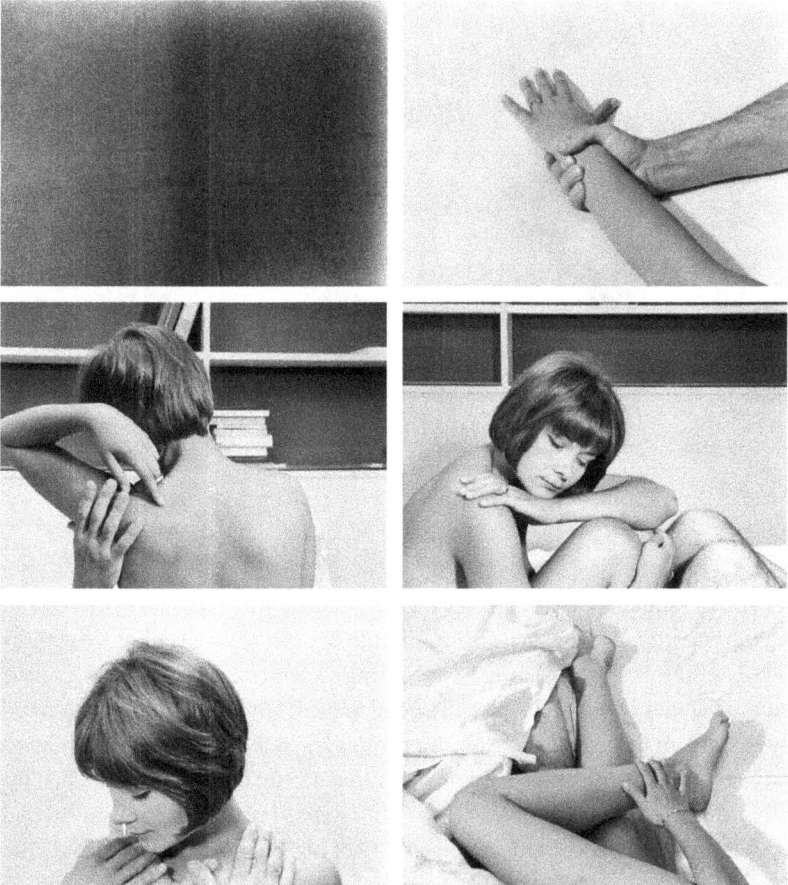

Fig. 5.9 *A Married Woman* (*Une femme mariée*, Jean-Luc Godard, 1964)

168 FACE AND HANDS

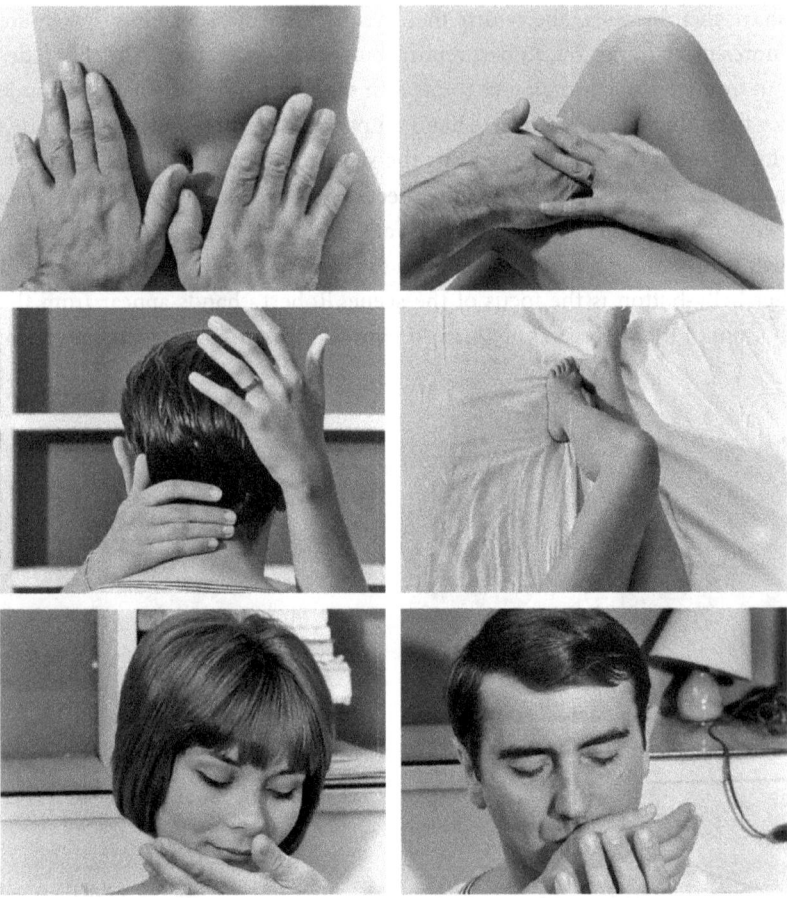

Fig. 5.9 Continued

Charlotte's hand appears and touches his for a moment. He says he wants to have her child. The shot fades. The seventh shot shows a detail of Charlotte's legs bent into a strange position; according to some cinema critics who have studied Godard's work, they are intended to form an A or a K;[55] Robert's hand appears from the left and caresses Charlotte's leg to the knee, reaches her hand and their fingers intertwine. Eighth shot: A close-up of Robert's nape, just for an instant we see Charlotte's face, then her hands start to caress his head and neck, while they talk about the possibility of her divorcing her husband. The

---

[55] Liandrat-Guigues and Leutrat (1994, p. 36).

ninth shot of this opening sequence focusses once again on Charlotte's legs on the bed, one slightly bent and the other stretched out; this time Robert's hands do not appear to disturb the harmony of the scene. The tenth shot shows an airplane in full flight, shattering just for a second the idyll of the two lovers but the eleventh is a continuation of the ninth and Robert's hands appear from the right. The twelfth shot is a close-up of Charlotte with Robert's hand touching her chin; she says she loves him. The next shot is specular: A close-up of Robert with Charlotte's hand touching his face; he says he loves her. The following three shots are brief close-ups of Charlotte, in profile, upside-down, and taken from below.

Any comment on these first striking 16 shots can apply to many other moments in this movie, which abounds in shots showing hands that grasp, caress, encircle, feet lifted up, close-ups of faces and objects, radios, billboards, isolated characters from letters, details of the clasp of a bra, and so on. As we have said before, this concept of a movie as a collation of fragments of life/living fragments is the very core of Godard's art, seeking the secret of contact in gestures, touching, brushing the skin with the fingertips; it is almost as if he were an entomologist looking at details through a microscope, hoping to find the sense of the relationship that the sequence of events of the movie intends to communicate as being vacuous and alienating. In an interview, the actress who played Charlotte, Macha Méril, compared Godard's shots to those filmed by an ornithologist.[56] In his lectures on cinema, Godard wrote of this movie:

> I didn't even realise I was trying to find it, this way of looking at the gesture. The same gesture, the fact of waiting and that you're going to catch something. Instead of catching a fish, you catch a hand, and this hand, depending on what it does to you, will also satisfy your hunger.[57]

Godard's comment regarding looking at a gesture seems to summarize quite effectively the notion of cinema based on action discussed in the previous chapter, although in this case the objective of the image is completely different. The way of looking at the gesture is constructed and conceived starting from its haptic component and transcends any possible form of judgment or inference that can give substance to a chain of actions able to move the story on from A to B. Jane Stadler would say that the haptic image denaturalizes the object of

---

[56] The interview is an extra on the Ripley's Home Video issue of *A Married Woman* (2011).
[57] Godard (1980, p. 128).

vision reducing the sense of depth and relying on the tactile perception of the texture of the image, which then takes precedence over form.[58]

The opening scenes of *The Married Woman* bring us face to face with cinema as the art of fragments, the exaltation of a form that has been fragmented and magnified, the carnal call exerted by surfaces (the human skin in this particular case). After just a few seconds of contemplation, Robert's hands come into play, doing exactly what the spectator would like to do, to still the hunger of which Godard speaks which is, of course, the impelling need to touch, the irresistible stimulus of contact that, according to Barker, is a constitutive part of our film experience. The spectator perceives much more than just body parts, he feels the carnality of the image, the richness of the corporeal content, where the substantiality of texture, the dense white and black fading into lighter shades, is present in all its superb materiality.

As we saw earlier, neuroscience offers a new interpretation of these images. Observing the contact between a hand and a face or a leg activates the corresponding tactile representation in the brain of the observer and triggers the simulation of those contacts through the reuse of somatosensory maps of SI, SII, and of the premotor cortex. In addition, Godard's body fragments show anatomic details blown out of scale, indeed almost out of the context of the bodies to which they belong, which probably potentiates the embodied simulation mechanism. Haptic visuality is therefore not merely a metaphor, it is also the description of a physiological process that connects the tactile sensations experienced by Charlotte and Robert with the tactile experiences of the spectator. Our hypothesis reintroduces the role of empathy in the experience of moving images, particularly images that show caresses, touch, intimate contact; by unconsciously simulating the content of these images in his brain-body, the spectator is able to establish an aesthetical relationship with them.

Animations

As early as the middle of the 1970s the great Czech filmmaker and artist Jan Švankmajer was working together with a group of clandestine Czech surrealists on tactile experiments in research on interpretation. They created objects and devices for collective experiments to gain understanding of the importance and the impact of touch when experiencing a work of art, as well as of the modifications generated by disgust following a tactile experience or

---

[58] Stadler (2008, p. 157).

simply the attention to this particular sense that is frequently underestimated in everyday life.[59] He had developed this research also for his puppet theater and later for his movies when he returned to animated film in the early 1980s. Švankmajer's interest in tactility in a medium that seemed to him supremely audiovisual was stimulated by Edgar Allan Poe. Reading and studying Poe, he discovered just how important touch was in the American author's imagination of horror and how these tactile sensations, even when transmitted by text with no direct bodily experience, were capable of having a particularly intense effect on tactile memory or tactile imagination. He was also greatly interested in Merleau-Ponty's *Phenomenology of Perception*, particularly the discussion of Kurt Goldstein's neurological cases that questioned the concept that touch is possible only in those cases in which there is direct contact with an object.[60] Many years before the experiments conducted by Marc Jeannerod, Goldstein wrote that when we are on the point of grasping a familiar object, even when that familiarity has only been established by tactile exploration and not by the actual sight of the object, the hand gradually assumes the form necessary to grasp it. It does this not only on the basis of the exterior form of the object, but also in response to a sort of central command: "We experience the sensation of a familiar object even before we grasp it," he wrote.[61] He anticipated by decades what neuroscience has recently demonstrated: Touch and movement are not only closely integrated, but can also be evoked by embodied simulation in the absence of explicit movements or contacts.

Švankmajer therefore decided to make a number of movies in order to try to "evoke tactile feelings, forgotten or buried, and to use them to enrich the emotional arsenal of methods of expression in my films," convinced of the innate sentiment of "emotional security" embodied in touch and exemplified, for instance, in the contact with the mother's body.[62]

The Fall of the House of Usher (*Zànik dom/Usher*, 1980) was the first extraordinary example of this concept of tactile cinema. Just as in the version directed by Jean Epstein in 1928, this story by Edgar Allan Poe underwent a visual adaptation that focussed principally on the sensitive and animated dimension of matter, which was after all Švankmajer's main line of research. Close-quarter images of muddy earth churned up by the hooves of invisible horses induce feelings of decomposition and abandonment in the spectator, decomposition and abandonment that are found again inside the house; Švankmajer attempts

---

[59] See Švankmajer and Švankmajer (1998, p. 85).
[60] See Švankmajer and Švankmajer (1998, p. 85).
[61] Goldstein (1939, p. 91).
[62] Švankmajer and Švankmajer (1998).

to provoke these feelings through tactile synesthesia, facilitated by the oozy and squelchy appearance of the earth (Figs. 5.10–5.12).

When Švankmajer uses details and close-ups, his purpose is not so much to create a referential relationship with the objects themselves, but with their tactile properties that immediately link up to the materiality of the texture of the images. When the camera enters the House of Usher, its gaze immediately skims over the mouldy walls, the chipped doors with their peeling paintwork abandoned over time, giving the spectator the sense of crumbling plaster, rippled wood, using almost informal abstract shots that we are accustomed to seeing today in meta-reflections on the decay of celluloid film, the spotting, swelling, and ripping that accompany its decay. The armchair positioned alone in the centre of the great hall is mainly an excuse to narrate the details of the backrest, filmed at such close quarters that it is not immediately apparent what it is (Fig. 5.11); outside the house once more, the spectator is impressed by Švankmajer's magistral shots of earth and water churned together to produce strange muddy forms (Fig. 5.12).

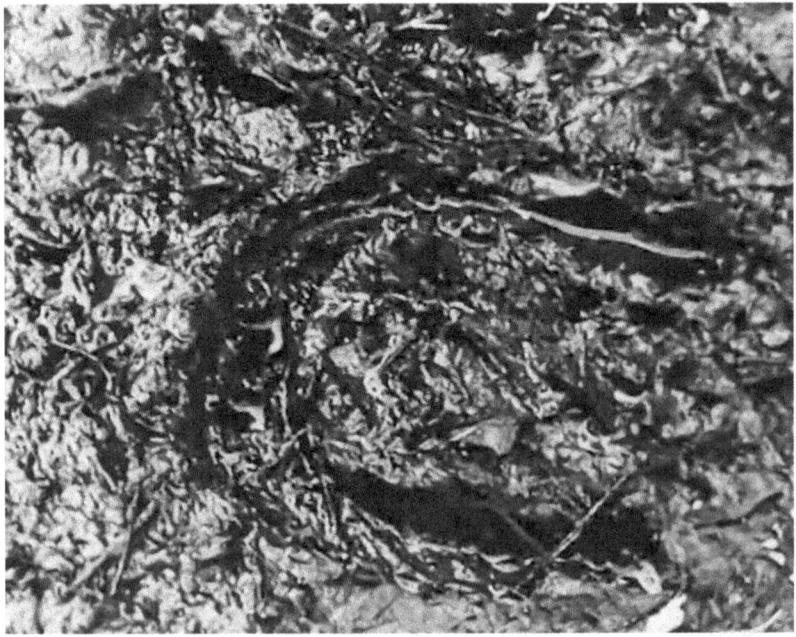

Fig. 5.10 *The Fall of the House of Usher* (*Zánik domů Usherů*, Jan Švankmajer, 1980).

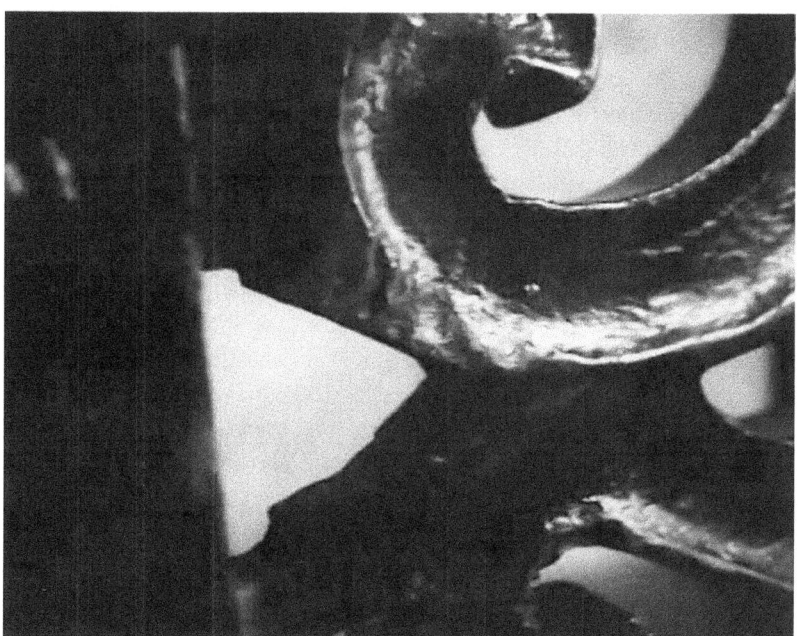

Fig. 5.11 *The Fall of the House of Usher* (*Zánik domü Usherü*, Jan Švankmajer, 1980).

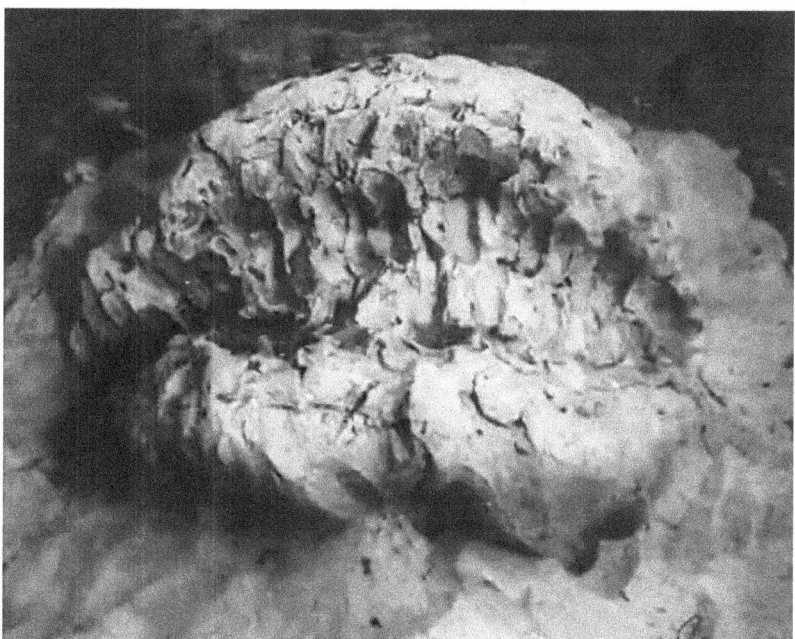

Fig. 5.12 *The Fall of the House of Usher* (*Zánik domü Usherü*, Jan Švankmajer, 1980).

He used this same technique in his famous movies with clay characters, such as *Dimensions of Dialogue* (*Možnosti dialogu*, 1982) and *Darkness Light Darkness* (*Tma/Světlo/Tma*, 1989). In *Dimensions of Dialogue,* we see a confrontation of odd characters, strange beings fashioned in Arcimboldo-style from various recycled objects, which finally disintegrate (or are disintegrated) and then recompose with different functions (Fig. 5.13).

During the process of auto-destruction, the food composing these golem-type creatures minces and churns, and Švankmajer shows us the resulting mess so close to that we have the feeling we can touch it, even smell it. Then he proposes a strange relationship between his classical clay figures, who size each other up, touch each other, participate in a strange form of decomposed intersubjectivity that causes them to liquefy into each other when they kiss (Fig. 5.14).

In *Darkness Light Darkness*, instead, the action takes place in an empty room (that brings to mind the famous Ames room), with two hands that collect limbs and organs and in so doing compose a human body; the scene of the clay fingers "wearing" two eyes that have slipped into the room and so seem to have acquired the gift of sight is particularly memorable (Fig. 5.15).

Fig. 5.13 *Dimensions of Dialogue* (*Možnosti dialogu*, Jan Švankmajer, 1982).

Fig. 5.14 *Dimensions of Dialogue* (*Možnosti dialogu*, Jan Švankmajer, 1982).

In both these movies, apart from the aesthetic aspects and the social ferocity that Švankmajer communicates, the focus is on imagination and tactile memory; the haptic visuality presupposed by the clay and the food is reinforced by the use of close-ups and magnification of the objects. The spectator is drawn to the objects, feels the layers of substance and how they are composed, and is led to reflect, once again, on the concrete structure of the film image. The extraordinary impression of tactile involvement that the spectator feels while watching these movies is to be attributed to simulations of his own tactile experiences due to the activation of the SI and SII areas and the premotor cortex. The effect is further amplified by the materiality of the clay figures, especially as clay is closely associated with contact manipulation and the creation of three-dimensional forms.

With these films, Švankmajer invites us to reflect on the haptic values of animation that represent an alternative to the values suggested by movies such as *Persona* or *A Married Woman*. In his book on the transformations that computer-generated imagery (CGI) has wrought in cinema, Dan North dedicates a section to the tactile nature of the digital image as one of the major concerns in the field of special effects, closely linked to the issue

176   FACE AND HANDS

Fig. 5.15 *Darkness Light Darkness* (*Tma/Světlo/Tma*, Jan Švankmajer, 1982).

of perceptual realism that the computer-generated image must be able to transmit. Referring to Christian Metz's considerations on tactility as the "supreme arbiter" of reality, he observes how important it is for those working in the post-production phase, fine-tuning digital images, to obtain just the right degree of shininess when dealing with metals or ice, of softness for fur, a realistic look of humidity or roughness when producing the skin of a Jurassic Park dinosaur.[63] The challenge lies in allowing the camera to "move in" on the details of the image created by the computer and to communicate a perceptual realism, sharpened by the tactile values of the virtually reproduced objects and bodies. Even in a case like this, as we will see shortly, close-ups and details play a key role on at least three fronts: They represent a testing ground for just how realistic the special effects and the synthespians (i.e., computer-generated actors) are; they modulate the distance of the spectator and his resonance vis-à-vis the surfaces of the objects and bodies he perceives; they create a bridge between the function of what is shown and the nature and texture of the digital image.

[63] North (2008, p. 21).

Fig. 5.16 *Toy Story* (John Lasseter, 1995).

In *The Tactile Eye*, Jennifer Barker analyzed two Pixar films from the haptic point of view, *Tin Toy* (1988) and *Toy Story* (1995), both directed by John Lasseter. Writing about *Toy Story*, which was a box office success that even appealed to an adult public, Barker wrote:

> *Toy Story* calls up its adult viewers' nostalgic impulses by appealing to the sensual childhood memories that reside at the surface of their skins. American audiences of a certain age will remember the feel of those little plastic green soldiers—the smooth flatness of the platforms to which their feet are fused and the intricate detail of their guns and helmets, for example, and the sharp edges they get when they've been chewed by the family dog. […] *Toy Story*'s slick computer-generated imagery not only accentuates the surface of its objects but creates for the film itself a skin that is without imperfection, at times hard and at times soft, but always as cool and smooth as a Fisher-Price toy.[64]

According to Barker, one of the reasons for the success of this movie—at least among adult viewers—is to be sought in the tactile memory that the new CGI is able to stimulate by reproducing toys. It is no coincidence that once the research studios of Pixar realized they were able to infuse plasticity and realism into games and toys, they worked, through the years, to regain ground in representing human characters. Although for many years the suggestive metaphor "skin of the film" seemed perfect for the celluloid strip, due to its impressionability and "warmth," with the advent of the computer it is now seen as suiting the "coldness" of the digital image. As Elsaesser and Hagener have pointed out, this numerical frigidity of the digital image concerns just the processes

---

[64] Barker (2009, p. 45).

of registration and filing that are the "dark side of its strengths," not the presentation and exhibition of the images: "its re-embodied manifestations of everything visible, tactile and sensory allow the digital to become much more closely aligned and attuned to the body and the senses."[65] So, Barker is justified in saying that *Toy Story* "does have a skin,"[66] a skin that provokes haptic and even erotic reactions in the spectator, which can produce a form of tactile engagement that is not devoid of memories and emotions that have their roots in a consumer culture so vividly represented by those smooth, Fisher-Price-style surfaces (Fig. 5.16). At the same time, the fact that *Toy Story* is digitally appealing obliges us to reflect on the nature of the digital image, on the new face of cinematography and the impact that the output of those numeric codes has on our brain–body system. Today, digital is the latest stage in the history of anthropomorphism applied to the technology of cinema, and as Elsaesser and Hagener have said, it brings back into play the ambitions of Dziga Vertov's "camera-eye", Béla Balàzs theory of the "visible man," Laura Marks' paradigm of the "skin of film," or Gilles Deleuze's "the screen as the brain."[67] Once again, the close-ups and the details of the toys, toys in the real sense of the word and not in Godard's de-humanized sense, facilitate an exploration of the new digital texture and our embodied relationship with it; all this stimulates us to investigate what kind of intersubjective and empathic relationship can be established with CGI.

Research on the synthespian is still in the initial stages, enthusiastically exploring the improvements in motion capture technology. This technology captures the gestures and facial expressions of flesh-and-blood actors, transferring them digitally onto virtual beings which then assume motions and expressivity aligned not only with the perceptual realism that is the holy grail of the CGI operator, but also with the affections, actions, and emotions of the spectator. Andrew Serkis has specialized in this particular form of acting and has "lent" his corporeality to Gollum of *The Lord of the Rings*, one of the most famous synthespians in the history of cinema. Serkis' mimicry was also used to animate the digital face of King Kong in the movie of the same name directed by Peter Jackson in 2005; although Serkis' facial expressions were used for only a quarter of Kong's facial animations, it is still true that the human agentivity instilled into the great gorilla as into Gollum was an essential factor in creating the realism of these characters and their ability to establish solid intersubjective

---

[65] Elsaesser and Hagener (2007, p. 198).
[66] Barker (2009, p. 46).
[67] Elsaesser and Hagener (2007, pp. 215–16).

relationships with the spectator (relationships that are much more complicated in films that did not have the benefit of this technology).[68]

With motion capture, the close-ups of King Kong and Gollum have the same value as a human face and the enhancement of digital aesthetics has ensured that the diaphanous skin of the monster of *The Lord of the Rings* and the thick fur of Kong maintain the haptic potential that many animators have aimed at over the years when reproducing the figments of their imagination. The haptic quality of Gollum's skin and Kong's fur, enhanced by digital technology and combined with the biological movements of the flesh-and-blood actor accomplished with motion capture technology, potentiate the embodied simulation of the movements on the screen and the tactile involvement mediated by the activation of the cerebral circuits in simulation mode that underlie the spectator's desire to touch that skin and fur. In fact, a recent fMRI study has shown that the mere intention to touch a body activates SI, which normally maps directly experienced tactile sensations.[69] According to our hypothesis, even in films produced with digital animation the empathic involvement we experience as spectators is the combined result of the simulation of the movements we observe and the potential contact of our hand with those digitally embodied surfaces.

---

[68] North (2008, p. 180).
[69] Ebisch et al. (2014).

6
New Mediation, New Films, New Experiments

In the preceding chapters, we examined the motor resonance that movies evoke in the spectator, interpreting it in the light of the theory of embodied simulation and the forms (and strengths) of cinema. In so doing we have reflected on the theories of the 1900s and analyzed silent cinema, the Hollywood classics, the new American cinema of the 1900s, European art cinema, experimental animation, and mainstream films. We have discussed the ways movies communicate with the spectator and the degree of involvement they manage to generate, but always in the light of what could be called a "traditional" filmic experience.

This final chapter will have a slightly different slant as we will be discussing how the filmic experience and films themselves have been transformed by the advent of digital media and the new forms of communication this has brought in its wake. As one of our objectives with this book is to examine the quality of the experience mediated by cinema and to evaluate to what degree embodied simulation supports the spectator's involvement in that experience, our mission would not be complete without an excursus into the realm of the new forms of mediation and how the spectator relates to them. And, of course, we need to see what impact, if any, the new technologies and devices for viewing, which are becoming increasingly popular, will have on our hypothesis of embodied simulation.

This chapter is therefore a bridge between the research that we have illustrated in the preceding pages and a study of new forms of mediation and films produced with the latest technology (and also, of course, of those new technologies that are capable of producing images with a cinematographic significance). By linking the past with the present, we will also have a glimpse of the experimental horizons and what the future may hold.

The Empathic Screen: Cinema and Neuroscience. Vittorio Gallese and Michele Guerra, Oxford University Press (2020). © Oxford University Press.
DOI: 10.1093/oso/9780198793533.001.0001

New positioning

One of the great revolutions that the digital era has wrought is the eclipse of traditional media apparatus and the convergence of social and artistic practices, which, although they may seem worlds apart, have been brought brusquely together by the new digital technology. We are writing this on our laptop, the same laptop we used a few hours ago to show students a PowerPoint presentation in which we had uploaded and edited a number of sequences taken from cinema classics, using the same laptop. Maybe a little later we will open the laptop again to watch our favorite TV series, look at the photos we took earlier in the day, Skype a friend, and we could even watch a movie on its small screen (although neither of us would really want to do this!).

What we have always thought of as film, experiencing movies and the community that gravitates around cinema has been drastically modified; according to some, damaged beyond repair, according to others simply transformed, and in many aspects enhanced, made more powerful. Of course, the same could be said for our social life and our network of relations. These days it would be difficult to find someone who would happily give up their laptop, tablet, or smartphone and live without the optionals that these devices offer. Vilém Flusser imagined a world where the telematic society would be like an ant colony, with each tiny ant in its own cell, connected by a "dream-secreting superbrain."[1] His impression was that the advent of this society would coincide with a process of "reduction," which would reflect a "tendency to the minute," physical and social bodies would progressively atrophy; in this telematic world, people will sit in their cells tapping keyboards with their fingertips in total isolation. The concept of enormity would acquire significance, paradoxically, for the very small, so that the infinitely large could be contained in the infinitely small, reversing the contemporary desire for growth and the aesthetics of the huge, making way for "less is more" and "small is beautiful." The world will be swallowed into telematic devices and reduced to a click, or even just to a fingertip swiping across a screen.

Many forms of convergence (the distribution of information on multiple platforms and the public's use of new research engines and forms of digital entertainment)[2] and remediation (the incorporation and representation of one medium in another)[3] would appear to confirm Flusser's view, offering evidence

---

[1] Flusser (1985, p. 131).
[2] Jenkins (2006, pp. 2–3).
[3] Bolter and Grusin (1999).

of a progressive reduction in the size of devices and increase in their portability. As Anne Friedberg pointed out, a century of cinema has accustomed us to the idea of a spectator sitting in front of a screen, mobility being guaranteed by the actions on that screen and the ability of those behind the structuring of this action to capture his gaze (and not only his gaze, as we have seen in this book). Although television transmitted different forms, content, and experiences, it inherited the concept of the "static spectator," who, of course, was free to choose what to see and when, but generally speaking did it from his armchair. Then came the twenty-first century with digital technology and the so-called new media, bringing with it a post-cinematographic and post-television concept of spectator, based on a substantial expansion of mobility that is no longer simply the virtual mobility of the spectator's gaze, but also the effective mobility of portable viewing devices with miniature screens that can be installed in trains and planes or be taken anywhere if equipped with wireless capabilities.[4]

Cinema has been significantly impacted by these changes: It has migrated into other technological devices, it has been "reduced" to laptop screens, to tablets, smartphones, even to those tiny screens set into headrests. Moving images are just exactly that; so faithful are they to their definition that they have glided out of their original habitat and are now everywhere, easily adapting to new rhythms and means of viewing. Francesco Casetti has dedicated his latest book to the new forms of cinema, providing an exhaustive overview of the media transformations that have swept over the "seventh art," while convincingly demonstrating that there continues to be "an idea of cinema" at the base of these new experiences even if they cannot be perceptually and socially associated with the traditional concept of the cinematographic experience.[5] According to Casetti, cinema is not reducing, cinema is expanding; it has invaded our homes (where we try to recreate the conditions of "going to the movies" by installing home theater technology in our sitting rooms) and, in fact, is now so mobile that we can shut out the real world and immerse ourselves in a movie wherever we are and whenever we feel so inclined. Of course, films viewed on small fluorescent rather than reflective screens, that are not the center of our attention and activity in the same way as they are when viewed at the cinema, cannot stimulate the same degree of pleasure in viewing that we have when we are at the movies; a pleasure that has alimented a cinephile mythology, and also much thought and study. However, there is much to be said for

---

[4] Friedberg (2006, p. 242). See Manovich (2001, pp. 128–49) for an overview of the concept of the screen and its relationship with the body of the user.
[5] Casetti (2015).

Casetti's position that this "relocation" cannot destroy our intimate concept of cinema that stimulates that feeling of anticipation every time we decide to go to the movies, nor can it destroy the cinematographic experience: "the concept of relocation makes clear that the migration of a medium outside its prior terrain involves a type of experience and a physical or technological space."[6]

Taking a look at the changes that "relocation" has brought with it, it is clear that, on the one hand, the post-cinematographic spectator must possess a degree of media competence (there is much more to watching a movie now than just going to the cinema, buying a ticket, finding your seat, and waiting for the movie to begin), and, on the other, there is the position the spectator must assume vis-à-vis the movie that now occupies physical and technological spaces he manages and even handles himself. The spectator's contact with the movie, which has been a *fil rouge* of this book, takes place very close to the screen (to watch a movie on a laptop or tablet with a level of involvement that will facilitate engagement, the spectator should ideally be close to the screen and move it when he changes his position in his chair) and frequently requires manipulation of the device. In this post-cinematic era, film is increasingly within hand's reach, or to be more precise, within "finger's reach." Casetti draws our attention to the deployment of new *sensorial dimensions*, in which the spectator actually (and visibly) activates sight and hearing, but also touch (and therefore movement)[7] in a very different way compared to the *haptic visuality* discussed in the preceding chapter. To put it briefly, whether the spectator is viewing a movie from a DVD drive, on a laptop, or on a smartphone, his hands will have an increasingly important role to play; as Casetti pointed out, in a way this is a return to the past, to the era of the pre-cinematographic devices of the 1800s such as the praxinoscope, the zoetrope, or the phénakistiscope, which to create the impression of movement necessitated an intervention by the spectator.

It is widely accepted, and a question of common sense, that viewing a movie on a digital screen is by no means the same as seeing it on the big screen. Whoever has had the experience of seeing a movie in a cinema and watching it on a laptop during a train journey or at home on their home theater will understand the basic differences, which to all extents and purposes are more closely related to the accessory experiences of cinema (getting out of the house, going to the cinema, seeing other people who have had the same idea, participation in what is to all effects and purposes a social event which will last as long as we remain in the cinema) and our disposition to immerse ourselves in the movie

---

[6] Casetti (2015, p. 29).
[7] Casetti (2015).

(the cinema commands total concentration, whereas with a laptop we can interrupt the viewing to prepare a drink, or to hypertext ... where did I see that actor before? Hold on a minute, I'll Google him ...).

In the second chapter we saw how in certain research studies on filmology, or on the anthropology and philosophy of cinema, the question of the spectator's position facing the screen is decisive for understanding the "miracle" of the dual experience in which he is physically glued to his seat but contemporaneously projected into the virtual space on the screen. The "cinematographic situation" was decisive for establishing the power of the cinema and guaranteeing this type of simulated mobility. It is generally thought that people who watch films by streaming or downloading them to their laptops pay for their greater media competence and power over the movie in terms of the loss of the magic the cinema exerts and a reduction of the coefficient of presence the filmic experience involves. At the end of the 1990s, when the changes mentioned earlier had not quite reached their apex but the direction they were taking became evident, the question of the "presence" (the spectator's physical and mental involvement in the plot) was reviewed and studied in the light of the decreasing dimensions of the screen. In 1999 Byron Reeves and his colleagues suggested that independently of the contents of the images and the type of scene, a large screen heightened the spectator's level of attention and arousal, albeit the former seemed to be weaker than the latter, which was particularly evident in scenes of sex or violence.[8] The literature on the topic always tended to support the idea that bigger is better,[9] but the arrival of portable devices gave rise to new studies and new hypotheses, which slowly but surely are challenging some of the more widespread assumptions regarding screen size. Cheryl Bracken and Gary Pettey, writing in 2002, sustained that the tiny screen of the iPod induces a greater sense of immersion in the spectator than do larger television screens.[10] The authors proposed three possible reasons for this, which remains the most sensational finding of the study: First, the novelty of the experience (several of the participants commented that they were not aware you could watch videos on iPods); second, as the participants were using earphones, their sense of immersion and isolation could have been heightened compared to watching the movie on television; last, but in our opinion not least, the fact of holding the iPod could enhance the sense of intimacy, inducing a physiological stimulation that should not be underestimated. Although researchers declare that there is

[8] Reeves et al. (1999).
[9] Silvera et al. (2002).
[10] Bracken, Pettey (2007).

still much to study, particularly regarding the interaction between the spectator and the visual contents of the movie, what has already been done certainly encourages new reflections on the sense of "presence" induced by smaller screens and the new devices that process moving images. Steven Bellman and his colleagues showed their participants TV advertising on screens of differing dimensions (laptop, smartphone, and iPod); their findings suggested that the distance from the screen, the type of shot, and the viewing angle are more important than the actual size of the screen for recall and degree of appeal.[11]

Andreas Baranowski and Heiko Hecht recently found that where the viewing takes place is more important than the dimension of the screen. These two researchers had their participants watch video clips in three paradigmatic conditions: In a real cinema; in a scale model of a cinema, built in a box reproducing the interior of a movie theater complete with real cinema seats and spectators, and a monitor at the end "mocked up" to resemble a screen; and lastly on a monitor with no modifications. The results showed a greater level of involvement when the video clips were seen in the real cinema and the model cinema. According to the authors, the results of the experiments we described earlier—in which a greater coefficient of immersion and presence was found for the smaller screens—are to be attributed to a lack of significant difference in the dimension of the screens used, as a real cinema screen can generate levels of immersion and presence that are significantly higher. This confirms what Tom Troscianko and colleagues found in their experiment with two screens of different dimensions onto which they projected the same sequence of Sergio Leone's *The Good, the Bad and the Ugly* that Uri Hasson used for his famous "neurocinematics" study.[12] Baranowski and Hecht maintain that their data support the "surround hypothesis," according to which presence and immersion are linked to the cinema itself and the presence of the public, which in the case of the model cinema can result in an overestimation of the size of the screen.[13] Regardless of how these results are argued, it would seem there is a general consensus that the cinema as an environment guarantees a stronger degree of presence and immersion (and therefore of simulation), having the advantage of the ambience, the sharing of the experience with other spectators, and the dimension of the screen. However, even when watching a movie in a different type of location and using a smaller screen, the spectator can still develop equally efficacious forms of simulation. It goes without saying that in this

---

[11] Bellman et al. (2009).
[12] Troscianko et al. (2012).
[13] Baranowski, Hecht (2014).

new digital experience the position and role of the spectator's body are completely different; however, it is to be expected that he will develop other competences and become accustomed to new forms of contact with the mediator (in this case, a screen built into a device). He will adapt, to be able to represent internally what he is seeing, so sustaining the ideas of authors such as Jonathan Gottschall, who sees human beings as story-telling animals,[14] or Frank Rose, who explores the future territories of the contemporary spectator's immersive experience, gliding from games to narrative, illusion and reality, marketing and information.[15] It is also conceivable that David Rodowick was right when he wrote that digital technology still has not fully accelerated the aesthetic transformation of the old analogical media, nor of a complete mutation of how they were experienced, so the spectator's vision and imagination is still supported by analogical models encapsulated in digital structures.[16]

Digital presences

We are convinced of the significance of the immersive potential of portable devices such as smartphones and tablets, which is still being discussed in the theoretic debate on cinema and the new media, and certainly merits further study. In our view, the type of immersion and involvement that these devices offer to the spectator is closely linked to the type of pragmatic relation he requires. Smartphones (and tablets to a lesser extent) are easy to handle, are located in our peri-personal space, require motor interaction (digitation to search for the movie, to select the viewing mode—streaming or download—the level of sound, presence of subtitles, fast forward, stop, and so on). True, many of these functions are available on normal remote controls, but as the name implies, they are, indeed, remote and function with devices that are situated at a certain distance from our body and decidedly external to the motor and spatial confines of our peri-personal space. This is a substantial difference in terms of proxemics vis-à-vis our relations with the device and probably transforms what we are watching into the potential object of a potential and possible interaction.[17]

---

[14] Gottschall (2012).
[15] Rose (2011).
[16] Rodowick (2007, p. 115).
[17] We must add that even the most traditional immersion supported by the simulation of the images observed on the screen can be shaped by the sight of actions shown within the spectator's peri-personal space. Mirror neurons are sensitive to the distance from which the scene is observed; a significant percentage responds only when observed close to the body of the observer, remaining inactive if the scene takes place at a distance. See Caggiano et al. (2009).

Today, interaction with portable digital devices goes far beyond simply pressing keys on a keyboard: It is now more a question of tactile contact with the image by means of constant contact with the screen. For the first time ever, the images we see on the screen are the result of an intentional motor act, as when we enlarge the image on the screen of our smartphone by swiping it with our fingers. If we want to see the image better by making it larger, we have to take deliberate action. As in a sort of retro-motion, the action mediated by the interactive screen is the key to its configuration (we can make the image larger or smaller at will by swiping our fingers across the screen). The image of the movie is "bound" to the action that has produced it by the screen being touched in a particular way. This results in a further welding of image and movement; in this case, however, the image not only evokes a mirroring of itself, but paves the way for a physical, almost carnal, relationship with the spectator.

It could be said that by means of a "real" haptic and sensory-motor stimulus generated by the active contact of our fingers with the surface of the screen, the images related to the narration of the movie we are watching *absorb* the haptic qualities of our control contacts, amplifying the normal corporeal involvement supported by the embodied simulation that is always present when we watch a movie, regardless of the medium.

We have seen how the diverse social contexts and relational styles that characterize the viewing of a movie on different media can influence and shape the level of involvement resulting from the presence evoked by the characters and the plot. This encourages us to investigate what changes, if any, the new "haptic era" of interactive portable digital devices will bring to the concepts of presence of images, of the immersion and involvement that they elicit.

According to Hans Gumbrecht, there are two components in aesthetic experience, one being meaning and the other presence.[18] Presence reflects the spectator's corporeal involvement through a synesthetic multimodal relationship with the artistic cultural artifact, the visual perception of which is qualified as "haptic vision". Gumbrecht sustains that it is possible to analyze every historical culture from the dual perspective of meaning and presence, as both are to be found in varying degrees in every cultural artifact. When presence predominates, artifacts acquire meaning from their intrinsic pragmatic sensory–motor inherence, and not from interpretation. Embodied simulation is a fundamental element of this relationship. As we have sustained throughout this book, our relations with the external world are not exclusively objective,

[18] Gumbrecht (2004).

from a third person perspective; we are part and parcel of that world, our body is an integral part and, at least partially, constitutes its origin.

While it is true that it is not possible to separate the experience of images from our daily motor, affective, tactile, and enteroceptive experience of reality, we believe that given the particular typologies of interaction of the new digital devices, they offer new opportunities for sensory–motor mediation of the image. Mark Hansen has written some very interesting articles on this theme.[19] In his view, the introduction of new digital technologies usurps the role of language as the predominant vector of reality, placing a new corporeal visuality at the center of our experience of the world. The technological modernization of the postmodern age has brought the body back into the center of our relationship with reality, a reality that is increasingly mediated by interactive digital visual representation. The increasing autonomy of the material digital world is insinuating itself more and more into our unconscious, contemporaneously externalizing our memories. The immediacy of the reality that we see every day on screens that follow us everywhere sneaks into our unconscious and takes it over, gradually, but relentlessly. From being linguistically reflective (*Erfahrung*), our experience of the world has become sensuously embodied (*Erlebnis*); according to Hansen, we have passed from the *semiosis* of reality to *mimesis*. He sustains that because of this, a new theorization that will account for the corporeal dimension of the impact digital technology has on our experience will have to be constructed. For Hansen, corporeal mimesis is the principal means for adapting to the technological mechanosphere of the postmodern age.

Unfortunately, we cannot linger on the multitude of stimuli that these considerations offer for our approach to the nature and power of filmic images. Suffice it to say that the new digital mediations are already changing correlatively the forms of artistic expression and reception. In this new "haptic age," images no longer have to rely on the narrative content for transmission to the spectator. As we saw in Chapter 5, long before the advent of the digital era cinema repeatedly used image texture and materiality to stimulate the spectator's senses. That said, the digital evolution has moved the axis of presence from the contents to the container, which then becomes the true attraction, not to mention object of desire, of the user (the term "spectator" is no longer appropriate in this context). In fact, the spectator-turned-user finds satisfaction in continuously manipulating his device, in the contact with the screen and the effect that his manipulations have on the images that float beneath its surface.

---

[19] Hansen (2004, 2006).

The hand is no longer only the mirror of the movements and the actions projected on the screen, as it was in the films traditionally viewed in extrapersonal space. In smartphones and tablets the images move within the peripersonal space that surrounds our body, *inside* the screen whose surface, like a transparent skin, circumscribes the world in movement that writhes beneath. By touching the surface of the screen with the tips of our fingers we establish a dual contact with these images. We modify them by touch, and they in turn touch us. The interactive digital screen probably represents the most extraordinary technological amplifier of the natural multimodality of viewing discussed at length in this book. We are convinced that Merleau-Ponty would have approved of these "chiasmatic" characteristics... [20]

Our corporeal interaction with the image means that it emanates appeal, the appeal of an object that we would wish to touch. This brings a new and technological form of "iconic act" (Horst Bredekamp's *Bildakt* or *Image-Act*)[21] in which the active role of the image is to offer itself as a physically tangible object of desire. Bredekamp asserted that Leonardo da Vinci was the first in modern times to attribute a central role in animating an image from the inside to the illusion of movement. For Leonardo, painting "is not alive, but only expressive of living things."[22] We are convinced that our interpretation of the perceptual effectuality of interactive digital images could be included in research studies on the underlying physiological and neural mechanisms; indeed, the content of the preceding chapters could provide a good starting point.

Death by chat

Unfriended (2014), directed by Levan (Leo) Gabriadze, gives an original twist to the *found-footage horror* sub-genre, but also represents a radicalization of the sub-genre as it deals with the more social aspects of the digital imaginary. These are films in which the spectator is brought face to face with the recording power of modern devices that have transformed our grip on reality, and that impassively oppose this power to violent crime and supernatural events. The fact that the surveillance regime in which we live (surveillance by the photo-cinematographic digital instruments built into practically all the devices we carry with us on a daily basis) retains everything, gives the authors of these

---

[20] To capture the new haptic dimension of film immersion that touch screens produce, we recently introduced the notion of 'skin-screen'. See Gallese and Guerra (2018); Gallese (2019).
[21] Bredekamp (2010).
[22] Bredekamp (2010, p. 99).

horror stories the opportunity of strengthening the impression of transparency and voyeurism typical of this type of movie.

The plot of *Unfriended* is centered on the suicide of Laura Barns (Heather Sossaman) that we watch, in all its dramatic realism, in a video posted on YouTube. The suicide is the consequence of another video posted on YouTube, showing Laura totally drunk and out of control at a party. We are not told who filmed the video or who posted it. A year after her death, her friends are chatting on Skype when a username unknown to them all appears in the chat. The account seems to be managed by Laura; whoever is behind it knows things that only Laura could know. She asks the participants in the group to confess who filmed the video that drove her to suicide. We will not reveal the details of the plot, we will just say that the mysterious force behind the account proceeds to eliminate her friends (violently) one by one, until only one person was left, the one person above suspicion, who naturally turned out to be the guilty party.

The entire movie is focussed on one single shot, a screen-cast of the desktop of one of the characters, Blaire (Shelley Hennig), on which we see windows showing the other participants in the chat (and into which, one after the other, horror will burst unannounced), the mysterious account managed by Laura, the threads of dialogues between the friends, and the opening of Facebook pages, YouTube, and so on, among which Blaire moves calmly at first, but gradually becoming panic-stricken as the movie unfolds. *Unfriended* includes contents of all types, iconic, linguistic, musical, photographic; at times they are knitted together according to the canons of the intersubjectivity and representation of the social networks, at others according to the virtual realism of our traditional use of the meta-medium computer. It can be said that this movie represents for convergence culture and remediation practices what Brian De Palma's *Redacted* (2007) represented for the various discursive and social forms that the moving image can assume. Pietro Montani's comments on *Redacted*, citing the movie as an example of contemporary intermediality, could also be applied, *mutatis mutandis*, to *Unfriended*. The facts of the movie, both past and present, take the form of "media *inscriptions*" sunk into aesthetic and social contexts that have been reconfigured by the functionality of the device containing them and rendering them available. In this case, too, and maybe even more strongly and more disturbingly, we recognize "the impersonal condition of filming oneself and being filmed, which seems to cover the entire world of action and suffering like a *media overwrap* without any apparent lacerations."[23]

---

[23] Montani (2010, pp. 64–5).

Like *Redacted*, *Unfriended* is a movie constructed with moving images that were not meant for movies.

But there is more to *Unfriended* than just this. Apart from the questions of the convergence and remediation, this little horror movie from Universal—which netted more than 15 times its production costs in its first weekend in cinemas in the USA—raises interesting relocation issues. Viewed at the cinema, *Unfriended* seems to suffer the potential of the big screen and cinematographic style. Paradoxically, to sharpen our presence vis-à-vis *Unfriended* and heighten the tension the movie is intended to induce, it is best seen on the neutral malleability of a laptop, renouncing the atmosphere of the cinema and the crowd, transforming our experience from communal to individual, as happens every time we shut ourselves off from the world with our laptop. The fact that the movie reproduces an experience which is familiar to many, particularly of the younger generation, evokes a simulation that increases if the atmosphere of the movie is reproduced in the appropriate viewing mode of the computer.

Unfriended has no camera movements, no editing. The movement in the movie is concentrated entirely on the characters in their tiny windows on the screen and the forms of interaction with their respective devices; one of them touches the laptop to adjust the shot, another moves the screen so as to see better, or to be better seen, Blaire decides to open windows on the screen that suddenly change the content of the shot. Gabriadze cannot afford to adopt classical solutions to intensify the tension as these would undermine the effect of perfect identity and continuity between the single shot of his movie and a real chat on Skype; he attempts to overcome this by replicating Skype glitches; these temporary interruptions in the signal produce an effect analogous to the classic black out, suggesting an impending crescendo of drama.

Just as the spectator/user is glued to his computer, so the agentivity of the characters is completely circumscribed to the laptop; their method of interaction is expressed in their physical contact with the object that links them together in a single virtual space (which gives the movie both a unit of time and a virtual unit of place). As when we are on Skype, and in many other social media contexts, our role as "interactor" is played out on several levels: In the first place we see the people with whom we are talking and with whom we are relating adhering to totally new rules of proxemics and spatiality which, however, are still based on traditional customs (no one would normally stick his face into the webcam while Skyping, generating a grotesque close-up); in the second place we see ourselves in a window on the screen, which would, of course, be impossible outside the virtual space of this particular relationship. This gives us the opportunity of verifying our position, adjusting our appearance, our role in that particular conversation; finally, we have the possibility of sharing information contemporaneously with, but external to, the direct contact with our interlocutor; we can maintain the acoustics while momentarily suspending visual contact whilst we write or do research.

Unfriended plays on this heightened interactivity, starting from the power that the resuscitated Laura has to intervene in the various conversations, post comments, and share information on the various social platforms, and also to erupt in the segregated spaces of the individual windows that appear on Blaire's desktop. As the movie evolves in its spiral of violence and death, the role of the characters changes gradually; from interactors they become defenceless spectators, unable to assist the friends they see to be in danger. They come face to face with the unreality of the intentions and actions they intended to exercise on the scene, tumbling into that form of "impotence" and spatial segregation that was typical of the "cinematographic era." The entire movie could be interpreted as a metaphor of the role that the individual inhabitants of the web and its social networks are convinced they have, of the illusory nature of this role, and the much stronger power of subjugation and potential annihilation that this seemingly neutral medium is able to wield.

Subjectivity in *Unfriended* is weakened by the Skype effect. For much of the movie the spectator behaves as a voyeur with access to the chat, almost as if he were another mysterious user; this impression is extremely strong in the opening sequences during which only Blaire and her boyfriend Mitch are online when Mitch asks Blaire to take off her blouse for him. The spectator doesn't really notice that practically the entire movie is filtered by Blaire's subjectivity (in many ways the movie is an experiment of the same genre as *Lady in the Lake*, which we examined in the second chapter of this book), because Blaire herself appears on the screen in her own personal window. This dissimulated subjectivity is revealed in all its force when the girl remains alone on the screen and particularly when the mysterious presence enters the "safety area" where Blaire is and in which the spectator is too, on the "other side" of the laptop which, dramatically, is suddenly snapped shut by an unknown hand.

It can be said that one of the main points of interest of *Unfriended* is the fact that it presupposes that the spectator will be using a laptop; a relocated spectator who is not going to be negatively affected by the downsizing of the movie from the big screen to the laptop screen, quite the opposite, because as a laptop user this downsizing will optimize his relationship with the story and the images of the movie.

This diegetic expedient, achieved by relocation, aligns the spectator and the characters almost to perfection. As the laptop used in the movie coincides with the laptop used to watch the movie, this alignment is heightened by events on the spectator's laptop and what is happening in the movie (as when watching a movie we hear the acoustic alerts that a message has arrived on our device *for us*, or a Skype call is waiting *for us*). This is an unusual situation in which embodied simulation creates disorientation rather than identification. This disorientation is caused by the uncertainty of where the boundary between fiction and reality lies and the spectator's constant need to redefine the boundary between himself and the screen, both of which are the result of the relocation. In films like *Unfriended* it is more than likely that embodied simulation contributes to creating an unsettling and almost psychotic situation for the spectator, as a result of this sensation of fluctuating boundaries.

It goes without saying that Laura Barns has a Facebook page, animated by those who consider that she is a piece of no good and others who have fallen totally in love with her.

"A new film grammar … "

A few years ago, at the Academy of Motion Picture Arts and Sciences, John Lasseter, one of the founding fathers of Pixar, explained that the future of cinema can be glimpsed in some of the most widely used portable or wearable devices such as smartphones and GoPro cameras. Speaking at a conference with the title "The New Audience: Moviegoing in a Connected World" (in which Henry Jenkins, the theoretician of the "convergence culture" also participated), Lasseter said: "The GoPro and the iPhone are here. They give a vibrancy you have never been able to have before. I think a new film grammar is going to come with these things."[24] What Lasseter is predicting, and he is by no means alone in this, is that devices created to reproduce images in movement intended for circuits other than cinema (it would not be correct to say "minor circuits" as some videos shot with smartphones and GoPros have greater visibility on YouTube and other similar platforms than certain films) are now ready to make their debut in mainstream cinema. Smartphones and GoPros have now reached levels of optical sophistication which would have been beyond imagination a few years ago, whilst the web now offers accessories such as dollies and Steadicam appliances for iPhone. The possibility of having a cinema studio in your pocket is coming increasingly closer to being reality, just as is the idea of further closing the gap between reality and film, satisfying

---

[24] Rainey (2015) includes excerpts from John Lasseter's speech.

our desire to view images and share them using technical devices to extend our vision (Google Glass), our limbs (smartphones), even our bodies (GoPros or action cams).

When the Italian philosopher Maurizio Ferraris decided to write an ontology of the cell phone in 2005, he certainly could not have imagined even in his wildest dreams that mobile videography would have burst onto the market just a few years later (he returned to the subject in 2010).[25] The advent of the cell phone, about which much was written in the early years of the new millennium from the point of view of the extended mind, opened a whole new discussion regarding talking, texting, recording, and spatially locating oneself and others, while the smartphone has taken this discussion even further, touching on the production and diffusion of images and the use of iconic contents that have reached previously-unheard-of levels of complexity for such a device. The new ontology of the cell phone is extremely complicated. Where once upon a time lighters were waved on high to express appreciation during concerts, now the darkness is studded with the light of thousands of smartphones as their owners try to capture the performance of the artists on the stage, with a view to sharing their experience with millions of others. Who can honestly say that they have never given in, at least once, to the temptation to use their smartphone to film something particularly worthy of being shared; a simple gesture on one of the most user-friendly devices in existence is all it takes to produce an amateur product with strong prosthetic connotations.

Evolution, however, is taking its toll of this connotation of amateurism. As we have seen, instruments are now available that improve the quality of videos taken by smartphone and aesthetics of mobile videography are now developing that are engaging on equal terms with theoretical discussions on cinema.[26] There are those who see the smartphone as a device that shortens the distance between the person who is filming and what is being filmed and achieves greater involvement of the "operator" rather than of the spectator. In one of his Free Newsreels, and precisely in Roma n. 1, entitled "Is cinema finished?", Cesare Zavattini said that cinema "always offers the shortest mediation," expressing his dream of an increasingly greater osmosis between the person filming, what is being filmed, and who will view the final product. He said that film "coincides" with the reality being filmed and with the position of

---

[25] Ferraris (2005); Ferraris and Terrone (2010, pp. 18–27). This latter publication appears as "*Il cinema nell'epoca del videofonino*" in *Bianco & Nero*, edited by Roger Odin (2010).
[26] See Odin (2010).

the operator within that reality: Let everyone take up their camera (read smartphone) and make their own movie.

The smartphone presupposes a very strong idea of subjectivity (described by Laurence Allard as "self-technology"), emphasized by the perceivable role of the hand and the arm that manipulate it.[27] Ferraris and Terrone, in an attempt to explain its mobility in the language Heidegger used in *Being and Times*, insist on *Zuhandenheit*, in other words, on its being ready to hand and easy to manipulate,[28] which takes us back to our reflections on the filmic image being ready to hand. The oscillations, blurriness, shots taken from unusual angles, improbable cropping, uncertainties, and acoustic intrusions are all part of a "videomobile" imaginary fairly codified and settled, inspired by the clear presence of a subjective enunciation "implanted" in the body of the operator himself. From this point of view, the dolly or the Steadicam for smartphones are a sort of neutralization of the first function and real potential of the smartphone (and the same can be said for the selfie stick, which also serves as a stabilizer and dolly for shooting videos). In other words, the video–photographic properties of the smartphone cannot be separated from the corporeality that gives them form, updating the subjectivity canons we have seen in action in the second chapter and the third, dedicated to camera movements.

The videophone radicalizes the relation according to which the forms of interaction used by film require the convergence of three distinct bodies: That of the spectator, that of the film itself, and, last, but not least, that of the operator; as we will see shortly, the same holds true also for action cams. The definition of corporeal image that David MacDougall gave in his book on film and ethnography comes in useful here, taking on additional pertinence vis-à-vis our position. MacDougall insists on the fact that these are images to be considered in a sense as "mirrors of our bodies, replicating the whole of the body's activity, with its physical movements, its shifting attention and its conflicting impulses toward law and order." He went on to say that these images "are not just images of other bodies, they are also images of the body behind the camera and its relations with the world."[29]

Traditional cinema only went so far as to suggest the existence of a body simulated by the camera, but it completely dissimulated the effective presence of a real body behind it. Now the situation is completely reversed and that body behind the camera marches into view, revising the role of operator/performer

---

[27] Allard (2010, p. 57).
[28] Ferraris and Terrone (2010, p. 18).
[29] MacDougall (2006, p. 3).

(albeit without any claim to professionality in that field). If we admit that Jean-Louis Schefer was correct in saying that the body of the spectator has always been the real focus of the filmic image,[30] today, in the era of portable devices, of cameras worn like jackets, augmented reality through Google Glass (check out the online video, *One Day*... which is a sort of *Lady in the Lake 2.0*),[31] this focus is on the body of the operator/performer. These new devices bring back into play themes linked to the corporeality of the mediated experience and relative forms of simulation, and certainly encourage further investigation of the medial construction of subjectivity, and of what appears to be a form of dialectics between the disembodiment and re-embodiment of our experiences.

However, it must be said that the advent of the GroPro and smartphone cinema of which Lasseter spoke in such glowing terms is taking a different turn and as often happens in digital transformations there is an attempt to replicate traditional experiences, renouncing or at least concealing its own specificity. An example of this is *Night Fishing* (*Paranmanjang*, 2012), the movie shot by the Korean filmmaker Park Chan-Wook and his brother Park Chan-Kyong with an iPhone 4. They used modified lenses and cinematographic equipment, which while leaving some fascinating elements attributable to the smartphone, greatly weakened the novelty. All the Park brothers actually managed to do was to show that it is, indeed, possible to shoot a movie with a smartphone, but the result is better if the device is "souped-up" with some cinematographic modifications... The same can be said for Robert Rodriguez' *Project Green Screen* (2013), developed with a Blackberry 10, while the movie by the Italian playwright and filmmaker Pippo Del Bono, *La Paura* (2009), fully exploited the scathing intrusiveness of the cell phone without touching up the images or modifying the device to give the end product a more "cinematographic" look. The very fact that a playwright who has always focussed on corporeality should accept the challenge of shooting a movie with a cell phone confirms certain of the considerations regarding the redefinition of the body and subjectivity that this device suggests. In her review of *La Paura*, Cristina Piccino wrote that "the search for a collective 'first person' in the images seems to have found the ideal means of expression: the cell phone that we use daily tones down the gesture of filming and opens the way for an uninterrupted monologue with reality that provokes reactions in others, reactions maybe of anger or disgust, but in which there will always be a grain of truth."[32]

---

[30] Schefer (1980, p. 110).
[31] The British TV series *Black Mirror* includes a special instalment on technological themes entitled *White Christmas*, which provides some interesting insights on the matter.
[32] Piccino (2009).

What remains to be seen is whether the smartphone will be able to introduce a new form of video aesthetics or if it will simply offer a less bulky and more flexible means of shooting scenes in the traditional cinematographic style.

What the smartphone still hasn't been able to do, however, seems well within the reach of other digital cameras: The action cams or GoPros, from the name of the firm that first put them on the market in 2004. These are small, compact videocameras with a wide-angle lens and a fixed focal distance with good field depth, extremely resistant to knocks and shocks. They can be attached to practically any part of the body or used in standalone mode. Their most distinguishing feature is that they can shoot scenes from unusual angles, and they are mainly used for shooting first-person films. Following their launch on the market, they enjoyed immediate and growing success, with sales doubling year after year. Almost 15 years later, it is estimated that they account for one-third of the small videocameras sold worldwide. A constantly growing community of users employ action cams to shoot films that they then upload to YouTube, Vimeo, and other similar platforms; many are spectacular and are practically always characterized by a first-person narrative. Videos shot with action cams represent a growing percentage of the one billion hours of video watched per day,[33] with 400 hours of videos being uploaded every minute. At the end of 2013, it was estimated that over 17 million videos shot with action cams had been uploaded to YouTube since the platform was created in 2005. Like smartphones, action cams are no longer exclusively employed by amateurs: The filmmaker Anthony Hemingway introduced scenes shot with an action cam in his movie *Red Tails*, devised and produced by George Lucas.

The introduction of this digital medium is favoring the development of a new, realistic style of audiovisual shooting, emphasizing the emotional components, particularly due to the embodiment that they stimulate in the viewer.[34] Ortiz and Moya have suggested that, as in documentaries filmed with handheld cameras, videos shot with action cams tend to a style with content that is less fictional and more sensational.

In GoPro films the action is shot alternating diverse visual angles, the images in the center of the scene are perfectly in focus, blurring in the background, and although they are basically stable they still communicate the operator's movements. The operator is frequently to be seen in the video, sometimes full figure but more often just his hands or feet. Ortiz and Moya are convinced that action cams finally provide the solution to shooting first-person videos, as they

---

[33] *Wall Street Journal*, February 27, 2017.
[34] Ortiz and Moya (2015).

are not affected by the technical difficulties that accompany the use of a bulky camera; they are even an improvement on the Steadicam as they reduce the distance between the operator and the object being filmed to just a few millimetres. Given that the operator is also the main character, the camera and the subject of the movie occupy practically the same spatial position, resulting in a perfect superimposition of the biological and the digital eye. In GoPro videos the body of the operator is the real focus of the image, even more so than in videos filmed with a smartphone.

Action cam videos can be divided into three main categories: The first, shot from a first-person perspective, basically excludes the body of the operator; the second centers on the operator himself rather than on his surroundings and is achieved by fixing the GoPro to a support attached to his body so that his every gesture and every emotion is captured on film; the third category embraces films in which the GoPro has been attached to an object (e.g., to skis or to a surfboard) external to the operator but which is one with his actions and movements. Near-to-perfect first-person films are now possible, as the action cam captures all the physical stress and movements of the operator that the Steadicam would have compensated or eliminated. It is highly likely that action cams will open up an entire new world for cinematography, paving the way for a style that will become increasingly more embodied, as its incidence on embodied simulation mechanisms increases. In these early days of their professional and cinematographic use, the impression is that a solution is needed to tone down the excesses and the adrenaline content of the first-person images in GoPro style, balancing them with a system that will help the spectator to find his way in this new and unpredictable visual space.

There can, however, be no doubt that cinema, which has already celebrated its first 120 years, will continue to embrace technology as it evolves, revealing the wonders of the world and astounding us with sights and sounds. After all, there are those who say that a return to sensoriality may well be the key to reinventing the cinema of the future...

Goodbye to the screen?

In June 2013, speaking at the opening ceremony for a new building for the School of Cinematic Arts of the University of California, George Lucas and Stephen Spielberg both predicted the possible implosion of the movie industry, given the increasing number of costly productions that have failed to appeal to the cinema-going public. Lucas spoke of a gradual migration of cinema to online platforms (which, indeed, is already happening) and the transformation of the remaining blockbusters into big-ticket events.

Spielberg went even further. He said that as long as we continue to stare at a rectangle of whatever size, be it a full-size cinema screen or a mini-tablet, we cannot talk of immersion. We've got to get rid of that, he said. We have to transport the spectator (player, in Spielberg's definition) into the heart of the experience so that wherever he turns, he is surrounded by a three-dimensional reality. This, for Spielberg, is the future: Reality and image converge, mediation disappears, the player is completely absorbed into the experience that has been created for him. Commenting on these predictions, Frank Rose pointed out that the screen has changed from being a "window" to being a "barrier"[35] and as such must be broken down. We have to go beyond the screen, rather like Buster Keaton did in *Sherlock Jr*. Rose runs through the evolution of the immersive practices that Spielberg wants to eliminate, maintaining that videogames have taken over the role of immersion-provider for the contemporary spectator/player and even goes so far as to imagine that Spielberg's fictional world is close to the virtual reality of the Oculus Rift type (he mentions the famous guillotine simulator in which the spectator, equipped with virtual reality (VR) glasses, finds himself on the scaffold with a screaming crowd surging around him and below him the basket into which his head will fall).

While Spielberg's vision of the future of cinema and the role of the spectator/player may seem futuristic, it is still possible to conjure up scenarios that are even more extreme. Thomas Nagel is generally remembered for his famous article published on *The Philosophical Review* in 1974 in which he asked "What is it like to be a bat?", concluding that knowing what it is like to be a bat is beyond our cognitive abilities.[36] In spite of all the information we have regarding the neurophysiology of the bat, we are not in a position mentally to feel what the bat feels when it flies through space, as we do not possess the sensory organs necessary to simulate this experience. In other words, Nagel's article brings us face to face with the fact that it is our brain–body system that establishes the perimeters to our mental imagination and simulation. Fascinating though this subject is, we have to limit ourselves to observing that the plasticity of the neurophysiological mechanisms of which we have written in this book and which mediate the reception of moving images (whether they are seen in a traditional cinema, on portable digital devices, or in VR) suggests that the possibility of entering into the subjective experience of another being, even when

---

[35] Rose (2013).
[36] Nagel (1974).

that being is so remote from our reality as the bat, is maybe not as far-fetched as was once thought.

We have explained how embodied simulation provides us with a mode of access to other beings and other worlds that is much more direct than that mediated by explicit language-based mentalization mechanisms. We have also shown that the relation modalities provided by embodied simulation do not apply to everything we see, particularly in those cases in which what we see is not a part of our behavioural repertoire and relative experiences. This is best illustrated by a functional magnetic resonance imaging (fMRI) experiment conducted by Buccino et al. in 2004;[37] they demonstrated that when their participants saw a dog barking, their motor systems did not activate. Why? Simply because we humans do not know how to bark, and, therefore, when we see a dog barking, our motor system does not resonate; we cannot simulate the action because we do not have the appropriate motor programs. It follows that we would not be able to simulate the flight of a bird, much less that of a bat. All the same, as Rochat and her colleagues demonstrated in an experiment using two monkeys after a period of training, the plasticity of the mirror mechanism permits the activation of the mirror neurons that respond to the execution and observation of a hand grasping an object, even when the action is performed with a mechanical utensil such as normal or inverted pliers.[38] They also showed that the mirror neurons discharged when the monkeys observed the experimenter spearing food with a sharp stick (the monkeys had been accustomed to seeing this being done but had never done it themselves nor been taught to do it). The mirror neurons responded to all these observed execution modalities, but with varying discharge times; when the monkeys were performing or observing actions they were familiar with or knew how to do themselves, the neurons were much more rapid in their response. When the monkey was watching the experimenter spearing the food, the mirror neurons were much slower in their response, which started only when the point of the stick pierced the food, continuing after this event. This means that the mirror neurons were able to generalize the goal of the action being observed, even when the modalities were known to the monkeys by sight only.

Now, just imagine watching, time after time, movements or actions filmed with a GoPro, movements or actions that are impossible for human beings to perform, such as the flight of a bird or bat. Are we sure that seeing these images time and time again would not give us the impression that they were part of

---

[37] Buccino et al. (2004).
[38] Rochat et al. (2010).

our repertoire? Could the gap between the bat's experience of flying and our own not be bridged by embodied simulation, using the generalization of motor programs with which we are familiar, supported also (if not above all) by the possibility of simulating these alien movements seen from a first-person perspective that previously was not available? The addition of appropriate tactile and proprioceptive stimuli of our body in simulated flight would further heighten the plausibility of the subjective experience of what we are observing. From there it is just a short step to hypothesizing an increase in the feeling of immersion, which would then reduce the experiential gap that separates us from experiencing "what it is like to be a bat" in flight ... or to resonate with the increasingly sophisticated forms of immersion, linked to the rapidly evolving versatility and creative power of the new technologies.[39]

Regardless of what the future may hold and what role screens will play in that future, there is no doubt that cinema will evolve as technology evolves; change is the name of the game and cinema will change, faithful to its mission of transporting the spectator into a special world, a world to which he can relate but which at the same time responds to his need to escape, to lose himself in realms of fantasy. A world with moving images that unfold his life story, his aspirations, his history just as those images in the Chauvet caves did, tens of thousands of years ago, where this story started.

---

[39] This new immersive use of images can obviously be put to manifold applications; for example, it could facilitate training of complex procedures or activities, such as surgery, or revive memories of long-forgotten experiences, such as crawling like babies. We are convinced that in this case neuroscience can provide inspiration and the key to interpreting new forms of mediation and ways of experiencing moving images of which we have not even started to dream.

Glossary

This glossary is not intended to be exhaustive, but simply to provide additional information for some of the terms used in the text.

Action cameras: Usually referred to as action cams, these are small videocameras with a fixed focal length wide angle lens and a good depth of field; they can be mounted anywhere, including on the operator. The principal characteristic of these cameras, which are mainly used for point-of-view (POV) shots, is that they can film footage from unusual viewpoints. They are also known as **GoPro cameras**, from the name of the company that first put them on the market in 2004.

Brain imaging: A set of various techniques, including EEG and fMRI, with the same objective of studying the brain's functional properties, measuring cerebral activity both directly and indirectly through the visualization of brain activity.

Continuity editing: This is a technical expedient used in editing to ensure that the transition from one shot to another is smooth and almost imperceptible. There are six fundamental edits, most of which are necessary to adhere to the so-called **180-degree rule**:

- Direction: When an actor exits a scene to the right, in the next scene he must enter to the left, to avoid the impression that he is retracing his steps. In this way, the two shots are linked by the trajectory of the actor's movement.
- Match on action: An action or a movement that started in one shot can be concluded in the next so that the two shots are linked by the continuity of the action or movement.
- Eyeline match: When the camera focusses on an actor whose gaze is directed off-screen (whether or not the spectator knows who or what the actor is looking at), the following scene typically shows what he was looking at, thus fulfilling the spectator's expectations.
- Shot/reverse shot is a technique that consists in alternating the **POV** of two actors who are standing or sitting face-to-face, occupying different areas of space. It is commonly used when shooting scenes with dialogues in which normally the two actors are filmed alternately, from angles that facilitate the spectator in orienting the action (actor x occupies the left-hand side of the shot and will direct his gaze off-screen to the right, while actor y occupies the right-hand side of the following shot and directs his gaze off-screen to the left).
- Sound: A sound originating on screen or from a source invisible to the spectator off-screen, or a line of dialogue, can be used to link two shots together, emphasizing their contiguity.
- Axial cuts: The same action can be shot at two different times and divided into two shots, of which the second can either seem to be much closer or further away but must always be on the same axis as the first.

Depth-of-field is an optical solution by which elements in a shot can be kept in focus (a large depth-of-field, also known as deep focus, is used to keep all the images sharp, while in a small depth-of-field, also known as shallow focus, some of the images will be slightly fuzzy).

EEG: The acronym for electroencephalography. This technique is used in **brain imaging** to directly record global electrical activity in the brain. It records the voltage fluctuations at the various frequencies (alpha, beta, gamma, delta, theta) produced by neuronal activity, not the activity of the individual neurons themselves. *ERD (event-related desynchronization)* can measure the modulations of these frequencies, induced by the coherent activities of numerous neuronal groups that although frequently located far apart, are still reciprocally connected. EEG can also be used to record *ERPs (event-related potentials)* that respond to different tasks (e.g., perceptive, cognitive, etc.). Its temporal resolution is optimum, in the millisecond range, but its spatial resolution is less satisfactory. Complex algorithms have to be used to locate the electrical events recorded by EEG; although their validity and reliability are good, they still require fine-tuning. Examples of how EEG can be applied to the reception of a movie are given in Chapters 3 and 4.

Embodied simulation: The verb "simulate" derives from the Latin "similis." It expresses the attempt to imitate the characteristics of a process or situation, using analogous means or strategies, with the objective of arriving at a better understanding, from the inside, so to speak. This concept has been applied to explain both motor control and our ability to understand others. Marc Jeannerod was one of the first to use the concept of simulation in neuroscience. In a powerful article published in 1994 he proposed that motor imagination could be considered as a form of simulation. When we imagine performing an action, certain physiological parameters behave exactly as if we were actually carrying it out and as the imaginary effort increases, so do our heart beats and breathing rhythm, just as they would if we were performing the action physically. Therefore, imagining a movement is the same as simulating it. This equivalence has also been demonstrated in the domain of vision.

Gallese and Goldman presented this relation between simulation, motor system, and mentalism in 1998. Starting with a description of the properties of mirror neurons, the authors suggested that there is a cognitive continuity between humans and the non-human primates in the ability to attribute intentional states to others, guaranteed by the ability, when observing others, to recognize the goals behind their actions. Gallese subsequently developed the hypothesis of embodied simulation, a functional architecture that constitutes a basic characteristic of the brain, relevant for social cognition. It is not limited exclusively to action but extends to other aspects of **intersubjectivity**. According to this hypothesis, the ability to understand the behavior of others and the underlying intentions, to imitate that behavior, to understand directly and experientially the sense of the emotions and sensations that others feel, depends on the constitution of a we-centric space, formed as a "shared manifold." This system can be defined operationally at three different levels: Phenomenological, functional, and sub-personal. The phenomenological level can be defined as empathic, a fundamental element for the experience of implicit certainties that we normally harbour vis-à-vis other people. Actions performed, sensations, and emotions experienced by others take on meaning for us due to the possibility of sharing them experientially through a common representational format. The functional level is represented by routine embodied simulation, "as-if" integration modalities that permit us to create models for ourselves and for others. The functional logic that lies at the basis of our control over our actions and experiences also operates during our implicit comprehension of the actions and experiences of other people. Both are expressions of *relational models*. Finally, the *sub-personal level* is constituted of the activity of a series of neural circuits, such as those of the canonical neurons, of mirror neurons, the neurons that map peri-personal space or those that map tactile sensations and emotions.

As has been explained throughout the book, the concept of embodied simulation coincides with the more corporeal aspects of empathy, but with a wider range of application that extends to the imagination and our relations with objects and the space around our body.

Embodiment: The notion of embodiment assigns the body a fundamental role in cognitive processes, emphasizing both the constitutive relationship between body, brain, and the world and the nature of corporeal representations that map the brain–world relationships by means of the body's mediations. The concept of embodiment is used in various acceptations that may be more or less radical depending on the importance attributed to the role of the physical body and/or that of the mental representations in bodily form; they hold in common the opinion that the study of the brain alone cannot explain our mental and cognitive activities. Content can be mapped or represented using different representational formats. For example, imagine that you want to give a friend directions on how to reach your home. You can do this in various ways: By physically showing him the route on a map, tracing it with your finger, or alternatively by sending him a map by e-mail—the content is the same, only the method of representation changes. It is therefore highly likely that what we call our mind also uses different formats, either linguistic or corporeal. The **embodied simulation** model implies the reuse of mental mechanisms and processes represented in corporeal form.

Empathy: You see someone accidentally hitting his thumb with a hammer and gasping in pain. You know he is in pain, but how do you know it? According to a classical cognitive version, this knowledge comes by means of a mental operation of "inference by analogy": In the past, you yourself have hit your thumb with a hammer and it hurt. Therefore, by analogy you infer that this other person, finding himself in the same situation, will experience the same sensation. According to the concept of empathy, however, you understand what the other person is feeling because while you are watching, you activate certain of the mechanisms that normally activate when you yourself feel pain, so you simulate *the other person's pain*. The concept of empathy was originally introduced in aesthetics in the second half of the nineteenth century to describe a particular type of corporeal involvement with objects, such as paintings and sculptures, that we contemplate from the artistic view point. It was subsequently extended to encompass a direct access modality to the world of others, which is the currently prevailing acceptation. Following the discovery of mirror neurons and other mirroring mechanisms, empathy is enjoying a new lease of life. The term actually means "feeling *with* another person" and it is important to distinguish it from the concept of sympathy, which is "feeling *for* another person." It is difficult to imagine experiencing sympathy without empathy, but not vice versa. Empathy is often used (erroneously) to designate other states such as emotional contagion, stepping into the "mental shoes" of another person, or sympathy itself. Through empathy you understand directly what another person is experiencing or doing, without losing the distinction between that other person and yourself.

fMRI is the acronym for *functional Magnetic Resonance Imaging*, a **brain imaging** technique that indirectly measures neuronal activity by recording the hemodynamic response, i.e., the variation of the level of oxygen in the local blood flow to the neurons. When neurons activate during a particular task, their metabolic activity increases, requiring a greater amount of blood, oxygen, and nutrients. fMRI measures the increase in blood flow as the difference in the local levels of oxygenized and deoxygenized haemoglobins, the so-called BOLD (*blood oxygen level-dependent*) signal. Using this technique, it is possible to estimate the blood flow variation in voxels with a temporal resolution of seconds. The results obtained are correlative, in that they correlate the increased activity in a given cerebral region with a specific task or function, without establishing a cause–effect relationship between the two.

GoPro cameras: See **Action cameras**.

Intersubjectivity: This term is used to denote the capacity of two or more people to share subjective states. It is one of the fundamental modalities for social cognition. The Theory of Mind and **empathy** are considered to be two modalities of intersubjectivity, with the former emphasizing the explicit deductive and inferential aspects of understanding others, whilst the latter focusses more on direct, implicit, and experiential aspects.

Jump cut: In editing, a jump cut is the juxtaposition of two sequential shots taken from opposing positions, so that the resulting effect is one of spatial–temporal discontinuity. It is a technique that flouts the classical rules of editing based on **continuity** shots or the **180-degree rule**.

Mirror mechanism is the term used for the functional property of various types of neurons that react to diverse modalities of the same stimulus. Actions, emotions, and sensations can induce the activation of groups of neurons both when they are performed or experienced directly in the first person and when their performance is observed in others. It is important not to confuse mirror mechanisms with **mirror neurons**; the latter only embody one of the aspects of the mirror mechanism in the motor system. In fact, the mirror mechanism takes on different functional characteristics and can contribute to understanding mental and cognitive processes depending on the species and the cerebral regions from which it is expressed.

Mirror neurons are multimodal motor neurons whose activation contributes to producing motor acts such as grasping an object with the hand or mouth. They also activate when a similar motor act is observed whilst being performed by another person, or even when the sound of this motor act being performed is heard. They also react when the most significant part of the motor act being observed is not visible and can only be imagined. Mirror neurons generalize their response to a given action, independently of who is performing it, or seeing or hearing it being performed. They were originally discovered in the premotor and parietal cortex of the macaque monkey, and subsequently neurons with the same properties were also found in birds, rats and humans.

GLOSSARY 207

Motion capture or *performance capture* is a technique that captures the body's natural movements and transcodes them digitally. Sensors are positioned on the actor's body or face to record movements that are then reproduced by a computer as a series of points that animators then map onto a model to obtain the impression of natural movement. The end product can be either an animated character whose gestures, movements, and facial expressions are deceptively human (excellent examples of this are to be found in Robert Zemeckis' films, such as *Polar Express, Beowulf,* and *A Christmas Carol*) or an imaginary creature such as Gollum in Peter Jackson's *The Lord of the Rings*. Jackson also used this technique to animate the great ape in *King Kong*, and in both these films the actor who "donated" his facial expressions and movements was Andrew Serkis, an expert in performance capture.

Neurons are the cells that make up the brain, or to be more precise the central and autonomous nervous system. One of their characteristics is excitability, they pass from a state of repose to a state of activity (the action potential or *spike*) in a very short period of time (approximately 1 millisecond). When they fire, neurons trigger many action potentials in rapid succession. If these are sufficiently intense and frequent, they activate or inhibit the next neuron with which they are in synaptic contact, so that it in turn fires or inhibits, depending on the type of synapsis. The synapsis is the structure that connects the neurons. The human brain contains approximately 100 billion neurons, each of which enters into synaptic communication with thousands of other neurons. Multiplying these figures, the result is astronomical. However, the complexity of the brain is not necessarily directly proportional to the number of its neurons and their multifarious connections.

Off-screen: See **Shot**

POV is the acronym for the *point-of-view shot* (see **Shot**).

Shot: The amount of space that the lens captures, coinciding with what the spectator sees. The space that is not captured by the lens is known as **off-screen**. When a character makes his entrance in a shot without there having been a jump cut, in film jargon he is said to "cross in." By varying the type of lens used and applying certain editing techniques, various spatial effects corresponding to the different shots can be obtained, for example:

- medium long shot (MLS), in which the actors are generally seen full figure, with the surroundings as background;
- long shot (LS), in which the surroundings are emphasized but the actors are still recognizable;
- extreme long shot (ELS) or extreme wide shot (EWS), in which the focus is on the surroundings, while the actors are seen in the distance and are rarely recognizable;
- full shot (FS) in which the entire area in which the action takes place is included in one shot.

Other types of shot are:

- **Establishing shot:** The term used to indicate a shot, usually the opening shot, that sets the context for a scene.
- **Over-the-shoulder shot:** This term is typically used when shooting dialogue scenes, framing one character from behind the shoulder of the other (see **semi-subjective** shot).

- **Point-of-view shot** (also known as a **subjective shot**): A shot which is taken in such a way as to suggest that it is exactly what the character is seeing. It is also referred to with its acronym **POV**.
- **Semi-subjective:** This is a variant of the **subjective** or **POV** shot, in which the character remains in the scene but is often shot from behind. Also known as the **over-the-shoulder shot**.

Steadicam: A harness or vest equipped with a system of shock absorbers, worn by the operator who can then move freely even over rough terrain without any oscillations or vibrations being visible in the movie. It was invented by Garrett Brown and made its debut in cinema in the mid-1970s.

The 180-degree rule: Usually the space used in cinematography is the 180-degree area in front of the camera. Imagine a classic dialogue scene set up for the **shot/reverse shot** technique; two characters are positioned one in front of the other, eye to eye, character x on the left and character y on the right. Now draw an imaginary line between the two that marks the 180-degree space, excluding the other 180-degree space beyond them. The camera focusses first on character x, then swings to character y and can move freely in the delimited 180-degree space (even changing the angulations of the shot/reverse shot), which allows the spectator to orient himself in the scene. If the camera were to cross this imaginary line, the spectator's perception would be distorted; character x would appear to be to the right instead of the left and character y would appear to be on the left. Even the points of reference in the background would be distorted and the cut between this shot and the next would be clearly perceptible. Sometimes the filmmaker deliberately moves the camera over the 180-degree line to create an effect of disorientation in the viewer.

TMS is the acronym of *transcranial magnetic stimulation*. A magnetic field generator is used to create a highly focalized magnetic field that can stimulate just a few millimetres of cerebral cortex. It is non-invasive; the distal part of the stimulator, known as the coil, is placed on the patient's scalp and can be moved to stimulate various regions of the brain. It can be used to establish a causal relation between the stimulation of a particular region and a given function. After a session of TMS, the patient may perform more rapidly and efficaciously. However, this treatment can also be used to deactivate a portion of the brain for the time it takes to verify what happens without the contribution of that particular area or to assess to what extent a person's performance on a particular task is affected when a given portion of the brain is deactivated.

Zoom: A lens with a focal distance that can vary in length, bringing the scene closer to or distancing it from the spectator. It simulates a dolly movement without actually moving the camera (and as mentioned in the text, it generates less relevant forms of motor resonance).

References

Adolphs, R., H. Damasio, D. Tranel, G. Cooper, A.R. Damasio (2000), "A role for somatosensory cortices in the visual recognition of emotion as revealed by three-dimensional lesion mapping", *Journal of Neuroscience*, 20, pp. 2683–2690.
Adolphs R., D. Tranel, A.R. Damasio (2003), "Dissociable neural systems for recognizing emotions", *Brain and Cognition*, 52, pp. 61–69.
Aglioti S.M., G. Berlucchi (2013), *Neurofobia*, Raffaello Cortina Editore, Milan.
Allard, L. (2010), "Mobilescape: il circuito delle immagini nella cultura portatile", *Bianco & Nero*, 568, pp. 56–63.
Allegri, L. (1976), *Ideologia e linguaggio nel cinema contemporaneo: Jean-Luc Godard*, CSAC-Università di Parma, Parma.
Alonge, G. (2012), *Scrivere per Hollywood. Ben Hecht e la sceneggiatura nel cinema americano classico*, Marsilio, Venice.
Alovisio, S. (2013), *L'occhio sensibile. Cinema e scienze della mente nell'Italia del primo Novecento*, Kaplan, Turin.
Amengual, B. (1971), *Clefs pour le cinéma*, Editions Seghers, Paris.
Ammaniti M., V. Gallese (2014), *The Birth of Intersubjectivity. Psychodynamics, Neurobiology and the Self*, W. W. Norton & Company, New York, p. 236.
Anderson, J.D. (1996), *The Reality of Illusion: An Ecological Approach to Cognitive Film Theory*, Southern Illinois University Press, Carbondale and Edwardsville.
Anderson, M.L. (2010), "Neural reuse: A fundamental reorganizing principle of the brain", *Behavioral Brain Sciences*, 33, pp. 245–266.
Angelini M., M. Calbi, A. Ferrari, B. Sbriscia-Fioretti, M. Franca, V. Gallese, M.A. Umiltà (2015), "Motor inihibition during overt and covert actions: An electrical neuroimaging study", *PLoS One*, 10, e0126800.
Angelucci, D. (2012), "Percepire una quasi-realtà. La fruizione cinematografica nell'estetica di Ingarden", in A. Pinotti (ed.), *Il piacere dell'opera*, Unicopli, Milan, pp. 90–98.
Arnheim, R. (1957), *Film as Art*, University of California Press, Berkeley and Los Angeles.
Artaud, A. (1976), *Antonin Artaud, Selected Writings*, Farrar, Straus, and Giroux, New York.
Auerbach, J. (2007), *Body Shots: Early Cinema Incarnations*, University of California Press, Berkeley and Los Angeles.
Auster, P. (2002), *The Book of Illusions*, New York, Henry Holt & Company.
Azéma, M., F. Rivère (2012), "Animation in Paleolithic art: A pre-echo of cinema", *Antiquity*, 86, pp. 316–324.
Balázs, B. (1924), "Visible Man", in "Early Film Theory: the Visible Man and The Spirit of Film", Berghahn Books, New York and Oxford, pp. 9–15.
Balázs, B. (1931), *Estetica del film*, tr. it. Editori Riuniti, Rome.
Ball, E.H. (1913), *The Art of the Photoplay*, G.W. Dillingham Company, New York.
Ball, E.H. (1915), *Photoplay Scenarios: How to Write and Sell Them*, Hearst's International Library Company, New York.
Banda D., J. Moure (text chosen and presented by) (2008), *Le cinéma: naissance d'un art. Premiers écrits (1895-1920)*, Flammarion, Paris.

Baranowski, A., H. Hecht (2014), "The big picture: Effects of surround on immersion and size perception", *Perception*, 43, pp. 1061–1070.
Barker, J. (2009), *The Tactile Eye: Touch and the Cinematic Experience*, University of California Press, Berkeley and Los Angeles.
Barratt, D. (2007), "Assessing the reality status of film: fiction or non-fiction, live action or CGI?", in J.D. Anderson, B. Fisher Anderson (eds.), *Narration and Spectatorship in Moving Image*, Cambridge Scholar Publishing, Newcastle, pp. 62–79.
Bazin, A. (1967, 2005), *What is cinema?* University of California Press, Berkely and Los Angeles, California.
Bellman, S., A. Schweda, D. Varan (2009), "Viewing angle matters—screen type does not", *Journal of Communication*, 59, pp. 609–634.
Bellour, R. (2009), *Le corps du cinema. Hypnoses, émotions, animalités*, POL, Paris.
Belton, J. (ed.) (1997), "Screenwriters and Screenwriting", *Film History*, 3.
Benjamin, W. (2005), *Selected Writings, Vol. 2, Part 1, 1927–1930*, Belknap Press of the Harvard University Press, Cambridge, MA, and London.
Bergman, I. (1959), "Chacun de mes films est le dernier", *Cahiers du cinéma*, 100, Octobre.
Bettman, G. (2013), *Directing the Camera: How Professional Directors Use a Moving Camera to Energize Their Films*, Michael Wiese Productions, Studio City, CA.
Blakemore, S. J., D. Bristow, G. Bird, C. Frith, J. Ward (2005), "Somatosensory activations during the observation of touch and a case of vision-touch synaesthesia", *Brain*, 128, pp. 1571–1583.
Bocher, M., R. Chisin, Y. Parag, N. Freedman, Y. Meir Weil, H. Lester, E. Mishani, O. Bonne (2001), "Cerebral activation associated with sexual arousal in response to a pornographic clip: A 15O-H2O PET study in heterosexual men". *NeuroImage*, 14, pp. 105–117.
Bolognini, N., A. Rossetti, S. Convento, G. Vallara (2013), "Understanding other's feelings: The role of the right primary somatosensory cortex in encoding the affective valence of other's touch", *Journal of Neuroscience*, 33, pp. 4201–4205.
Bolter, J.D., R. Grusin, (1999), *Remediation. Understanding New Media*, MIT Press, Cambridge, MA.
Bordwell, D. (1977), "Camera movement and cinematic space", *Ciné-Tracts*, 2, pp. 19–25.
Bordwell, D. (1998), *On the History of Film Style*, Harvard University Press, Cambridge, MA.
Bordwell, D. (2006), *The Way Hollywood Tells It: Story and Style in Modern Movies*, University of California Press, Berkeley and Los Angeles.
Bordwell, D. (2013), "The viewer's share: models of mind in explaining film", in A.P. Shimamura (ed.), *Psychocinematics: Exploring Cognition at the Movies*, Oxford University Press, New York, pp. 46–50.
Bordwell, D., N. Carroll (eds.) (1996), *Post-Theory: Reconstructing Film Studies*, The University of Wisconsin Press, Madison.
Bordwell, D., K. Thompson (2011), *Minding Movies: Observations on the Art, Craft, and Business of Filmmaking*, The University of Chicago Press, Chicago.
Bordwell, D., J. Staiger, K. Thompson (1985), *The Classical Hollywood Cinema: Film Style and Mode of Production to 1960*, Columbia University Press, New York.
Botvinick, M., A. P. Jha, L. M. Bylsma, S. A. Fabian, P. E. Solomon, K. M. Prkachin (2005), "Viewing facial expressions of pain engages cortical areas involved in the direct experience of pain", *NeuroImage* 25, pp. 315–319.
Bracken, C.C., G. Pettey (2007), "It is REALLY a smaller (and smaller) world: Presence and small screens", in L. Moreno (ed.), *Proceedings of the Tenth Annual International Meeting of the Presence Workshop*, Barcelona, pp. 283–290.
Branigan, E. (1992), *Narrative Comprehension and Film*, Routledge, London.

Branigan, E. (2006), *Projecting a Camera: Language-Games in Film Theory*, Routledge, New York.
Bredekamp, H. (2010), *Image Acts: A Systematic Approach to Visual Agency*, Walter De Gruyter GmbH, Berlin/Boston.
Bremmer F., A. Schlack, N.J. Shah, O. Zafiris, M. Kubischi, K. Hoffman, K. Zilles, G.R. Fink (2001), "Polymodal motion processing in posterior parietal and premotor cortex: A human fmri study strongly implies equivalencies between humans and monkeys", *Neuron*, 29, pp. 287-296.
Brown, T. (2012), *Breaking the Fourth Wall: Direct Address in the Cinema*, Edinburgh University Press, Edinburgh.
Bruno, M.W. (1999), *Tutti i film di Stanley Kubrick*, Gremese, Rome.
Buccheri, V. (2010), *Lo stile cinematografico*, Carocci, Rome.
Buccino, G., F. Lui, N. Canessa, I. Patteri, G. Lagravinese, F. Benuzzi, C.A. Porro, G. Rizzolatti (2004), "Neural circuits involved in the recognition of actions performed by nonconspecifics: An fMRI study", *Journal of Cognitive Neuroscience*, 16, pp. 114-126.
Bufalari, I., T. Aprile, A. Avenanti, S.M. Alioti (2007), "Empathy for pain and touch in the human somatosensory cortex", *Cerebral Cortex*, 17, pp. 2553-2561.
Burr, D. (2005) "Vision: In the blink of an eye", *Current Biology*, 15, pp. 554-556.
Caggiano V., L. Fogassi, G. Rizzolatti, P. Thier, A. Casile (2009), "Mirror neurons differentially encode the peripersonal and extrapersonal space of monkeys", *Science*, 324, pp. 403-406.
Caggiano, V., L. Fogassi, G. Rizzolatti, J.P. Pomper, P. Thier, M.A. Giese, A. Casile, A. (2011), "View-based encoding of actions in mirror neurons of area F5 in macaque premotor cortex". *Current Biology*, 21, 144-148.
Calbi, M., K. Heimann, D. Barratt, F. Siri, M.A. Umiltà, V. Gallese (2017), "How context influences our perception of emotional faces: A behavioral study on the Kuleshov effect", *Frontiers in Psychology*, 8, p. 1684.
Calbi, M., F. Siri, K. Heimann, D. Barratt, V. Gallese, A. Kolesnikova, M.A. Umiltà (2019), "How context influences the interpretation of facial expressions: A source localization high-density EEG study on the 'Kuleshov effect'", *Scientific Reports*, 14, p. 2107.
Calder A.J., J. Keane, F. Manes, N. Antoun, A.W. Young (2000), "Impaired recognition and experience of disgust following brain injury", *Nature Neuroscience*, 3, pp. 1077-1078.
Carbone, M. (2011), *La chair des images: Merleau-Ponty entre peinture et cinéma*, Vrin, Paris.
Carluccio, G. (2006), "Questioni di stile", in P. Bertetto (ed.), *Metodologie di analisi del film*, Laterza, Rome-Bari, pp. 107-146.
Carr, L., M. Iacoboni, M.C. Dubeau, J.C. Mazziotta, G.L. Lenzi (2003), "Neural mechanisms of empathy in humans: A relay from neural systems for imitation to limbic areas", *Proceedings of the National Academy of Sciences*, 100, pp. 5497-5502.
Carrillo, M., Y. Han, F. Migliorati, M. Liu, V. Gazzola, C. Keysers, (2019). Emotional Mirror Neurons in the Rat's Anterior Cingulate Cortex. *Current Biology*, 29, 1-12.
Carroll, N. (2007), "Narrative closure", *Philosophical Studies*, 135, pp. 1-15.
Casebier, A. (1991), *Film and Phenomenology: Toward a Realist Theory of Cinematic Representation*, Cambridge University Press, New York.
Caselli, R.J. (1991), "Rediscovering tactile agnosia", *Mayo Clinic Proceedings*, 66, pp. 129-142.
Casetti, F. (1998), *Inside the gaze*, Indiana University Press, Bloomington, Indiana.
Casetti, F. (1999), *Theories of cinema, 1945-1990*, University of Texas Press, Austin, Texas.
Casetti, F. (2008), *The eye of the century*, Columbia University Press, New York
Casetti, F. (2009), "Filmic experience", *Screen*, 50, pp. 56-66.

Casetti, F. (2015), *The Lumière galaxy*, Columbia University Press, New York.
Chateau, D. (2011), *La subjectivité au cinéma. Representations filmiques du subjectif*, Presses Universitaires de Rennes, Rennes.
Ciment, M. (1980), *Kubrick: The Definitive Edition*, Faber&Faber, Londonc.
Coëgnarts, M., P. Kravanja (2012), "Embodied visual meaning: Image schemas in film", *Projections*, 2, pp. 84–101.
Coëgnarts, M., P. Kravanja (2014), "A study in cinematic subjectivity: Metaphors of perception in film", *Metaphor and the Social World*, 2, pp. 149–173.
Coëgnarts, M., P. Kravanja (eds). (2015), *Embodied Cognition and Cinema*, Leuven University Press, Leuven.
Comuntzis, G.M. (1987) "Children's comprehension of changing viewpoints in visual presentations", Visual Communication Conference.
Comuntzis, G.M., G. Page (1991) "Perspective-taking theory: Shifting views from Sesame Street", Visual Communication Conference.
Coplan, A., P. Goldie (eds.) (2011), *Empathy: Philosophical and Psychological Perspectives*, Oxford University Press, New York.
Cosmides, L., J. Tooby (1997), "The multimodular nature of human intelligence", in A. Schiebel, J. W. Schopfds (eds.) *Origin and Evolution of Intelligence*, Center for the Study of the Evolution and Origin of Life, UCLA, Los Angeles, pp. 71–101.
Costa, A. (2008), "Introduzione", in V. Lindsay (ed.), *L'arte del film*, tr.it. Marsilio, Venice.
Craig, A.D. (2002), "How do you feel? Interoception: The sense of the physiological condition of the body", *Nature Reviews Neuroscience*, 3, 500–505.
Currie, G. (1995), *Image and Mind. Film, Philosophy, and Cognitive Science*, Cambridge University Press, Cambridge.
Currie, G. (2011), "Empathy for objects", in A. Coplan and P. Goldie (eds.) *Empathy: Philosophical and Psychological Perspectives*, Oxford University Press, New York, pp. 82–98.
Cutting, J.E., K.L. Brunick, A. Candan (2012), "Perceiving event dynamics and parsing Hollywood films", *Journal of Experimental Psychology: Human Perception and Performance*, 38, pp. 1476–1490.
Dagrada, E. (2015), *Between the Eye and the World: The Emergence of the Point-of-View Shot*, Peter Lang, Berlin.
Dalmasso, A.C. (2013), "Voir selon l'écran. Autour d'une rencontre entre visibilité et théorie filmique", in M. Carbone (under the direction of), *L'empreinte du visuel. Merleau-Ponty et les images aujourd'hui*, MetisPresses, Geneva, pp. 107–125.
D'Aloia, A. (2013), *La vertigine e il volo. L'esperienza filmica fra estetica e neuroscienze cognitive*, Ente dello Spettacolo, Rome.
D'Aloia, A., R. Eugeni (2015), "Neurofilmology: An introduction", *Cinéma et Cie*, 22–23, pp. 9–26.
Damasio, A.R. (1994), *Descartes' Error: Emotion, Reason, and the Human Brain*, G.P. Putnam's Sons, New York.
Damasio, A.R. (1999), *The Feeling of What Happens: Body, Emotion and the Making of Consciousness*, Harcourt Brace, San Diego.
Damasio, A.R. (2008), "Cinéma, esprit, émotion: la perspective du cerveau", *Trafic*, 67, pp. 94–101.
Damasio, A.R. (2010), *Self comes to Mind: Constructing the Conscious Brain*, Pantheon Books, New York.
Damasio, A.R., G.B. Carvalho (2013), "The nature of feelings: Evolutionary and neurobiological origins", *Nature Reviews Neuroscience*, 14, pp. 143–152.

De Amicis, E. (1907), *Cinematografo cerebrale*, Salerno Editrice, Rome, 1995.
Decety, J., H. Sjoholm, E. Ryding, G. Stenberg, D. Ingvar, (1990), "The cerebellum participates in cognitive activity: Tomographic measurements of regional cerebral blood flow", *Brain Research*, 535, pp. 313-317.
De Gaetano, R. (2013), *La potenza delle immagini. Il cinema, la forma, le forze*, ETS, Pisa.
Deleuze, G. (1983), *Cinema 1. The Movement Image*, Continuum, New York, 1986.
Deleuze, G. (1986), "Le cerveau, c'est l'écran. Entretien avec Gilles Deleuze", *Cahiers du cinéma*, 380, pp. 25-32.
Derambure, P., L. Defebvre, K. Dujardin, J.L. Bourriez, J.M. Jacquesson, A. Destee, J.D. Guieu (1993), "Abnormal cortical activation during planning of voluntary movement in patient with epilepsy with focal motor seizures: event-related desynchronization study of electroencephalographic mu rhythm", *Epilepsia*, 38, pp. 655-62.
Diamond, M.R., J. Ross, C. Morrone (2000), "Extraretinal control of saccadic suppression", *Journal of Neuroscience*, 20, pp. 3449-3455.
Di Pellegrino G., L. Fadiga, L. Fogassi, V. Gallese, G. Rizzolatti (1992), "Understanding motor events: A neurophysiological study", *Experimental Brain Research*, 91, pp. 176-180.
Dudai, Y. (2008), "Enslaving central executives: Toward a brain theory of cinema", *Projections*, 2.
Ebisch, S. J., M.G. Perrucci, A. Ferretti, C. Del Gratta, G.L. Romani, V. Gallese (2008), "The sense of touch: Embodied simulation in a visuotactile mirroring mechanism for observed animate or inanimate touch", *Journal of Cognitive Neuroscience*, 20, pp. 1611-1623.
Ebisch, S.J., F. Ferri, A. Salone, M.G. Perrucci, L. D'Amico, F.M. Ferro, G.L. Romani, V. Gallese (2011), "Differential involvement of somatosensory and interoceptive cortices during the observation of affective touch", *Journal of Cognitive Neuroscience*, 23, pp. 1808-1822.
Ebisch, S.J., F. Ferri, G.L. Romani, V. Gallesse (2014), "Reach out and touch someone: anticipatory neural representations of active interpersonal touch", *Journal of Cognitive Neuroscience*, 9, pp. 2171-2185.
Ejzenštejn, S.M. (1986), *Il montaggio*, in the care of P. Montani, Marsilio, Venice.
Ejzenštejn, S.M. (1995), "History of the close-up", in R. Taylor (ed.) *Beyond the Stars: The Memoirs of Sergei Eisenstein, Selected Works*, British Film Institute, London, vol. IV, pp. 461-478.
Ejzenštejn, S.M. (2009), *Sulla biomeccanica. Azione scenica e movimento*, under the care of A. Cervini (ed.), Armando Editore, Rome.
Elsaesser, T., M. Hagener (2007), *Film Theory: An Introduction Through the Senses*, Routledge, Taylor and Francis Group, New York and London.
Emerson, J., A. Loos (1921), *Breaking Into the Movies*, George W. Jacobs & C., Philadelphia.
Epstein, J. (2002), *L'essenza del cinema. Scritti sulla settima arte*, tr. it. under the care of V. Pasquali (ed.), Fondazione Scuola Nazionale di Cinema, Rome.
Eugeni, R. (1995), *Invito al cinema di Stanley Kubrick*, Mursia, Milan, 2014.
Falkenstein, M., J. Hoormann, S. Christ, J. Hohnsbein (2000), "ERP components on reaction errors and their functional significance: a tutorial", *Biological Psychology*, 51, pp. 87-107.
Farah, M.J. (1989), "The neural basis of mental imagery." *Trends in Neuroscience*, 12, pp. 395-399.
Farah, M.J. (2000), "The neural bases of mental imagery", in M.S. Gazzaniga (ed.) *The Cognitive Neurosciences* 3rd ed., The MIT Press, Cambridge, MA, pp. 975-986.
Farassino, A. (1974), *Jean-Luc Godard*, Il Castoro, Milan.
Feldmann, E. (1956), "Considerations sur la situation du spectateur au cinéma", *Revue Internationale de Filmologie*, 26, pp. 83-98.

Ferrara, S. (2001), *Steadicam: Techniques and Aesthetics*, Focal Press, Oxford.
Ferrari P.F., V. Gallese, G. Rizzolatti, L. Fogassi (2003), "Mirror neurons responding to the observation of ingestive and communicative mouth actions in the monkey ventral premotor cortex", *European Journal of Neuroscience*, 17, pp. 1703-1714.
Ferraris, M. (2005), *Dove sei? Ontologia del telefonino*, Bompiani, Milan.
Ferraris, M., E. Terrone (2010), "Doppia firma. Ontologia del mobile movie", *Bianco & Nero*, 568, pp. 18-27.
Flanagan, O. (1992), *Consciousness Reconsidered*, The MIT Press, Cambridge, MA.
Flusser, V. (2011), *Into the Universe of Technical Images*, University of Minnesota Press, Minneapolis.
Fodor, J. (1975), *The Language of Thought*, Thomas Y. Crowell Company, New York.
Fodor, J. (1983), *The Modularity of Mind*. The MIT Press, Cambridge, MA.
Fox, P., J. Pardo, S. Petersen, M. Raichle (1987), "Supplementary motor and premotor responses to actual and imagined hand movements with positron emission tomography", *Society for Neuroscience Abstracts* 13, p. 1433.
Freeburg, V.O. (1918), *The Art of Photoplay Making*, The MacMillan Company, New York.
Freeburg, V.O. (1923), *Pictorial Beauty on the Screen*, Arno Press & The New York Times, New York.
Freedberg D., V. Gallese (2007), "Motion, emotion and empathy in esthetic experience", *Trends in Cognitive Sciences*, 11, pp. 197-203.
Friedberg, A. (2006), *The Virtual Window: From Alberti to Microsoft*, The MIT Press, Cambridge, MA.
Friederici, A.D., E. Pfeiffer, A. Hahne (1993), "Event-related brain potentials during natural speech processing: effects of semantic, morphological and syntactic violations", *Cognitive Brain Research*, 1, pp. 183-192.
Friederici, A.D. (2002), "Towards a neural basis of auditory sentence processing", *Trends in Cognitive Sciences*, 6, pp. 78-84.
Furman, O., N. Dorfman, U. Hasson, L. Davachi, Y. Dudai (2007), "They saw a movie: Long-term memory for an extended audiovisual narrative", *Learning & Memory*, 6, pp. 457-467.
Gallese, V. (2000), "The inner sense of action: agency and motor representations", *Journal of Consciousness Studies*, 7, pp. 23-40.
Gallese, V. (2001), "The 'Shared Manifold' hypothesis: From mirror neurons to empathy", Journal of Consciousness Studies, 8, pp. 5-7, 33-50.
Gallese, V. (2003), "The manifold nature of interpersonal relations: The quest for a common mechanism", *Philosophical Transactions of the Royal Society of London B*, 358, pp. 517-528.
Gallese, V. (2005), "Embodied simulation: from neurons to phenomenal experience", *Phenomenology and the Cognitive Sciences*, 4, pp. 23-48.
Gallese, V. (2006), "Intentional attunement: A neurophysiological perspective on social cognition and its disruption in Autism", *Brain Research Cognitive Brain Research*, 1079, pp. 15-24.
Gallese, V. (2007), "Before and below Theory of Mind: Embodied simulation and the neural correlates of social cognition", *Philosophical Transactions of the Royal Society of London B*, 362, pp. 659-669.
Gallese, V. (2008), "Mirror neurons and the social nature of language: The neural exploitation hypothesis", *Social Neuroscience*, 3, pp. 317-333.
Gallese, V. (2011), "Neuroscience and phenomenology", *Phenomenology & Mind*, 1, pp. 33-48.
Gallese, V. (2012), "Aby Warburg and the dialogue among aesthetics, biology and physiology", *Ph*, 2, 48-62.

Gallese, V. (2014), "Bodily selves in relation: Embodied simulation as second-person perspective on intersubjectivity", *Philosophical Transactions of the Royal Society of London B Biological Sciences*, 369, pp. 20130177.
Gallese V. (2015), "Bodily Framing", in C. Jones, R. Uchill, D. Mather (eds.), *Experience: Culture, Cognition and the Common Sense*, The MIT Press, Cambridge, MA, pp. 237-248.
Gallese V. (2019), "Naturalizing aesthetic experience. The role of (liberated) embodied simulation". *Projections*, 12(2), pp. 50-59.
Gallese V., V. Cuccio (2015), "The paradigmatic body. Embodied simulation, intersubjectivity and the bodily self", in T. Metzinger, J.M. Windt (eds.), *Open MIND*, Frankfurt, pp. 1-23.
Gallese, V., S. Ebisch (2013), "Embodied simulation and touch: The sense of touch in social cognition", *Pnenomenology & Mind*, 4, pp. 269-291.
Gallese, V., M. Guerra (2012), "Embodying movies: Embodied simulation and film studies", *Cinema: Journal of Philosophy and the Moving Image*, 3, pp. 183-210.
Gallese, V., M. Guerra (2013a), "Film, corpo, cervello: prospettive naturalistiche per la teoria del film", *Fata Morgana*, 20, pp. 77-91.
Gallese, V., M. Guerra (2013b), "Forme di simulazione e sti(mo)li cinematografici", *Reti, Saperi, Linguaggi*, 2.
Gallese, V., M. Guerra (2014), "Corpo a corpo: simulazione incarnata e naturalizzazione dell'esperienza filmica", *Psicobiettivo*, 1, pp. 156-177.
Gallese, V., M. Guerra (2015), "The feeling of motion: Camera movements and motor cognition", *Cinéma & Cie*, 22-23, pp. 103-112.
Gallese, V., M. Guerra (2018), "L'empathie d'une machine", in M. Carbone, J. Bodini, A.C. Dalmasso (sous la direction de), *Des pouvoirs des écrans*, Mimesis International, Paris
Gallese, V., G. Lakoff (2005), "The brain's concepts: The role of the sensory-motor system in reason and language", *Cognitive Neuropsychology*, 22, pp. 455-479.
Gallese, V., C. Sinigaglia (2008), "What is so special with embodied simulation", *Trends in Cognitive Sciences*, 15, pp. 512-519.
Gallese, V., Sinigaglia, C. (2011a), "What is so special with embodied simulation?", *Trends in Cognitive Sciences*, 15, pp. 512-519.
Gallese, V., C. Sinigaglia (2011b), "How the body in action shapes the self", *Journal of Consciousness Study*, 7-8, pp. 117-143.
Gallese, V., C. Keysers, G. Rizzolatti (2004), "A unifying view of the basis of social cognition", *Trends in Cognitive Sciences*, 8, pp. 396-403.
Gallese, V., L. Fadiga, L. Fogassi, G. Rizzolatti (1996), "Action recognition in the premotor cortex", *Brain*, 119, pp. 593-609.
Gallese, V., M. Rochat, G. Cossu, C. Sinigaglia (2009), "Motor cognition and its role in the phylogeny and ontogeny of intentional understanding", *Developmental Psychology*, 45, pp. 103-113.
Gallese, V., Gernsbacher, M.A., Heyes, C., Hickock, G., Iacoboni, M. (2011), "Mirror neuron forum", *Perspectives on Psychological Science*, 6, pp. 369-347.
Gaudreault, A. (2008), *Cinéma et attractions*, CNR, Paris.
Geuens, J.P. (1993-1994), "Visuality and power: The work of the Steadicam", in *Film Quarterly*, 42.
Ghazanfar, A.A., S.V. Shepherd (2011), "Monkeys at the movies: What evolutionary cinematics tells us about film", *Projections*, 2, pp. 1-25.
Ghezzi, E. (1995), *Stanley Kubrick*, Il Castoro, Milan.

Gibson, J.J. (1979), *The Ecological Approach to Visual Perception*, Classic Edition, Psychology Press, New York, 1986.

Glenberg A. and V. Gallese (2012), "Action-based language: A theory of language acquisition production and comprehension", *Cortex*, 48, 905-922.

Godard, J.L. (1980), *Introduction to a True History of Cinema and Television*, Caboose, Montreal, 2012.

Godard, J.L. (1985), *Due o tre cose che so di me. Scritti e conversazioni sul cinema*, tr.it. under the care of O. Leogrande (ed.), Minimum Fax, Rome.

Goldman, A. (2012), "A moderate approach to embodied cognitive science", *Review of Philosophy and Psychology*, 3, pp. 71-88.

Goldman, A., Gallese, V. (2000), "Reply to Schulkin," *Trends in Cognitive Sciences*, 4, pp. 255-256.

Goldman A., F. De Vignemont (2009), "Is social cognition embodied?", *Trends in Cognitive Science* 13, pp. 154-159.

Goldstein, K. (1939), *The organism: A holistic approach to Visual Perception*, Zone Books, New York.

Goodman, N. (1975), "The Status of Style", *Critical Inquiry*, 4, pp. 799-811.

Gottschall, J. (2012), *The Storytelling Animal: How Stories Make Us Human*, Houghton Mifflin Harcourt Publishing, New York, 2013.

Grodal, T. (2009), *Embodied Visions, Evolution, Emotion, Culture and Film*, Oxford University Press Inc., New York.

Grodzinsky, Y., A. Santi (2008), "The Battle for Broca's Region", *Trends in Cogniitve Sciences*, 12, pp. 474-480.

Guerra, M. (2007), *Il meccanismo indifferente. La concezione della Storia nel cinema di Stanley Kubrick*, Aracne, Rome.

Guerra, M. (2013), "A young and enthusiastic champion of the cinema: Victor Oscar Freeburg e la natura del film", in V.O. Freeburg (ed.), *L'arte di fare film*, tr. it. under the care of M. Guerra (ed.), Diabasis, Parma, pp. 7-40

Guerra, M. (2014), "La teoria Americana degli anni Dieci: 'Action, action, ACTION!'", *Fata Morgana*, 23, pp. 27-35.

Guerra, M. (2015a), "Dall'elogio all'elegia. Ambizioni e posizioni della teoria del cinema", *Fata Morgana*, 26.

Guerra, M. (2015b), *La coscienza del film*, in U. Cocconi, G. Miranda, M. Pesenti Gritti (eds.), *Il primato della coscienza*, Diabasis, Parma, pp. 151-165.

Guerra, M. (2015c), "Modes of action at the movies: Or re-thinking film style from the embodied perspective", in M. Coëgnarts, P. Kravanja (eds.), *Embodied Cognition and Cinema*, Leuven University Press, Leuven.

Guerra, M. (2019), "Barry Lyndon", in E. Carocci (ed.), *Stanley Kubrick*, Marsilio, Venice.

Gumbrecht, H.U. (2004), *Production of Presence: What Meaning Cannot Convey*, Stanford University Press, Stanford.

Gunning, T. (1990), "The cinema of attractions: Early film, its spectator and the avant-garde", in T. Elsaesser (ed.), *Early Cinema: Space, Frame, Narrative*, BFI, London, pp. 56-62.

Hahne, A., A.D. Friederici (1999), "Electrophysiological evidence for two steps in syntactic analysis. Early automatic and late control processes", *Journal of Cognitive Neuroscience*, 11, pp. 194-205.

Haneke, M. (2010), "Terror and utopia of form. Robert Bresson's *Au hasard Balthazar*", in R. Grundmann (ed.), *A Companion to Michael Haneke*, London, Wiley-Blackwell, pp. 565-574.

Hannon, W.M. (1915), *The Photodrama: Its Place Among the Fine Arts*, The Ruskin Press, New Orleans.
Hansen, M. (2004), *New Philosophy for New Media*, The MIT Press, Cambridge, MA.
Hansen, M. (2006), *Bodies in Code. Interfaces with Digital Media*, Routledge, New York.
Hansen, M. (2011), *Cinema and Experience. Siegfried Kracauer, Walter Benjamin, and Theodor W. Adorno*, University of California Press, Berkeley and Los Angeles.
Hasson, U., Y. Nir, I. Levy, G. Fuhrmann, R. Malach (2004), "Intersubject synchronization of cortical activity during natural vision", *Science*, 303, pp. 1634–1640.
Hasson, U., O. Landesman, B. Knappmeyer, I. Vallines, N. Rubin, D.J. Heeger (2008), "Neurocinematics: The neuroscience of film", *Projections*, 1, pp. 1–26.
Heidegger, M. (1927), *Being and Time*, Blackwell Publishing, Oxford, 1962.
Heimann, K. (2015), *How Movies Move Us Just the Right Way. Exploring the Role of Camera Movements and Montage in Human Film Perception*, Ph.D. Thesis, Department of Neuroscience, University of Parma.
Heimann, K., M.A. Umiltà, M. Guerra, V. Gallese (2014), "Moving mirrors: A high-density EEG study investigating the effect of camera movements on motor cortex activation during action observation", *Journal of Cognitive Neuroscience*, 26, pp. 2087–2101.
Heimann, K.S., S. Uithol, M. Calbi, M. A. Umiltà, M. Guerra, V. Gallese (2017), "Cuts in action: A high-density EEG study investigating the neural correlates of different editing techniques in film", *Cognitive Science*, 41, pp. 1555–1588.
Heimann, K., Uithol, S., Calbi, M., Umiltà, M.A., Guerra, M., Fingerhut, J., Gallese, V. (2019), "Embodying the camera: an EEG study on the effect of camera movements on film spectators' sensorimotor cortex activation", *PLoS One*, 14, pp. e01211026.
Hickok, G. (2014), *The Myth of Mirror Neurons. The Real Neuroscience of Communication and Cognition*. W.W. Norton & Co., New York.
Højbjerg, L. (2014), "The circular camera movement: style, narration, and embodiment", *Projections*, 2, pp. 71–88.
Hurley, S. (2002), *Consciousness in Action*, Harvard University Press, Cambridge, MA.
Husserl, E. (1931), *Cartesian Meditations. An Introduction to Phenomenology*. Springer, Berlin (1977).
Husserl, E. (1952), *Ideas. General Introduction to Pure Phenomenology*, Routledge, New York, 2002.
Hutchison, W.D., K.D. Davis, A.M. Lozano, R.R. Tasker, J.O. Dostrovsky (1999), "Pain related neurons in the human cingulate cortex", *Nature Neuroscience*, 2, pp. 403–405.
Ihde, D. (1979), *Experimental Phenomenology: Multistabilities*, State University of New York Press, Albany, 2012.
Ihde, D. (2002), *Bodies in Technologies*, University of Minnesota Press, Minneapolis.
Ihde, D. (2010), *Embodied Technics*, Automatic Press/VIP, Copenhagen.
Ishida, H., L.C. Grandi, L. Fornia, M.A. Umiltà, V. Gallese (2013), "Somato-motor haptic processing in posterior inner perisylvian region (SII/pIC) of the macaque monkey" *PLoS One*, 8, e69931.
Iwase, M., Y. Ouchi, H. Okada, C. Yokoyama, S. Nobezawa, E. Yoshikawa, H. Tsukada, M. Takeda, K. Yamashita, K. Yamaguti, H. Kuratsune, A. Shimuzu, Y. Watanabe (2002), "Neural substrates of human facial expression of pleasant emotions induced by comic films: A PET study", *NeuroImage*, 17, pp. 758–768.
Jabbi, M., J. Bastiaansen, C. Keysers (2008), "A common anterior insula representation of disgust observation, experience and imagination shows divergent functional connectivity pathways", *PLoS One*, 3, e2939.

Jackson, P. L., A. N. Meltzoff, J. Decety. (2005), "How do we perceive the pain of others: A window into the neural processes involved in empathy", *NeuroImage*, 24, pp. 771-779.

Järveläinen, J., M. Schürmann, S. Avikainen, R. Hari, (2001), "Stronger reactivity of the human primary motor cortex during observation of live rather than video motor acts", *Neuroreport*, 12, pp. 3493-3495.

Jeannerod M. (1994), "The representing brain: Neural correlates of motor intention and imagery", *Behavioral Brain Sciences*, 17, pp. 187-245.

Jeannerod, M., M.A. Arbib, G. Rizzolatti, H. Sakata (1995), "Grasping objects: The cortical mechanisms of visuomotor transformation", *Trends in Neuroscience*, 18, pp. 314-320.

Jenkins, H. (2006), *Convergence Culture. Where Old and New Media Collide*, New York University Press, New York.

Johnson, M. (2007), *The Meaning of the Body: Aesthetics of Human Understandings*, The University of Chicago Press, Chicago.

Jonas, H. (1966), *Organism and Freedom. An Essay in Philosophical Biology*. DFG and Universität Siegen, Siegen.

Jullier, L. (2002), *Cinéma et cognition*, L'Harmattan, Paris.

Keysers, C., B. Wicker, V. Gazzola, J. Anton, L. Fogassi, V. Gallese (2004), "A touching sight—SII-PV activation during the observation and experience of touch", *Neuron*, 42, pp. 335-346.

Kepley, V., Jr. (1986), "The Kuleshov Workshop", *Iris*, 1, pp. 5-23.

Kohler, E., C. Keysers, M.A. Umiltà, L. Fogassi, V. Gallese, G. Rizzolatti (2002), "Hearing sounds, understanding actions: action representation in mirror neurons", *Science*, 297, pp. 846-848.

Koivisto, M., A. Revonsuo (2003), "An ERP study of change detection, change blindness, and visual awareness", *Psychophysiology*, 40, pp. 423-429.

Kosslyn, S. M. (1994), *Image and Brain: The Resolution of the Imagery Debate*. The MIT Press, Cambridge, MA.

Kosslyn, S. M., N. M. Alpert, W. L. Thompson, V. Maljkovic, S. Weise, C. Chabris, S. E. Hamilton, S. L. Rauch, F. S. Buonanno (1993), "Visual mental imagery activates topographically organized visual cortex: PET investigations", *Journal of Cognitive Neuroscience*, 5, pp. 263-287.

Kracauer, S. (1960), *Theory of Film: The Redemption of Physical Reality*, Princeton University Press, Princeton.

Kuehn E., R. Trampel, K. Mueller, R. Turner, S. Schütz-Bosbach (2013), "Judging roughness by sight—a 7-Tesla fMRI study on responsivity of the primary somatosensory cortex during observed touch of self and others", *Human Brain Mapping*, 34, pp. 1882-1895.

Lageira, J. (2011), "Imaginary subject", in D. Chateau (ed.), *Subjectivity: Filmic Representation and the Spectator's Experience*, Amsterdam University Press, Amsterdam, pp. 150-151

Lakoff, G. (1987), *Women, Fire and Dangerous Things*, University of Chicago Press, Chicago.

Lakoff, G., M. Johnson (1980), *Metaphors We Live By*, University of Chicago Press, Chicago.

Lakoff, G., M. Johnson (1999), *Philosophy in the Flesh: The Embodied Mind and its Challenge to Western Thought*, Basic Books, New York.

Le Bihan, D., R. Turner, T. A. Zeffiro, C. A. Cuénod, P. Jezzard, V. Bonnerot (1993), "Activation of Human Primary Visual Cortex During Visual Recall: A Magnetic Resonance Imaging Study", *Proceedings of the National Academy of Sciences*, 90, pp. 11802-11805.

Legrenzi, P., C. Umiltà (2009), *Neuro-mania. Il cervello non spiega chi siamo*, Il Mulino, Bologna.

Levin, D.T., A.M. Hymel, L. Baker (2013), "Belief, desire, action, and other stuff: Theory of mind in movies", in A.P. Shimamura (ed.), *Psychocinematics. Exploring Cognition at the Movies*, Oxford University Press, New York, pp. 244-266.
Liandrat-Guigues, S., J.L. Leutrat (1994), *Godard. Alla ricerca dell'arte perduta*, tr.it. Le Mani, Recco (GE).
Lo Piparo, F., (2003) *Aristotele e il linguaggio*, Laterza, Rome-Bari.
MacDougall, D. (2006), *The Corporeal Image: Film, Ethnography, and the Senses*, Princeton University Press, Princeton.
McGinn, C. (2005) *The Power of Movies: How Screen and Mind Interact*, Vintage Books, New York.
Maffongelli, L., E. Bartoli, D. Sammler, S. Koelsch, C. Campus, E. Olivier, L. Fadiga, A. D'Ausilio (2015), "Distinct brain signatures of content and structure violation during action sequence observation", *Neuropsychologia*, 75, pp. 30-39.
Magliano, J.P., J.M. Zacks (2011), "The Impact of Continuity Editing in Narrative Film on Event Segmentation", *Cognitive Science*, pp. 1-29.
Manovich, L. (2001), *The Language of New Media*, MIT Press, Cambridge, MA.
Marcus, L. (2007), *The Tenth Muse: Writing About Cinema in the Modernist Period*, Oxford University Press, New York.
Marks, L.U. (2000), *The Skin of the Film: Intercultural Cinema, Embodiment, and the Senses*, Duke University Press, Durham.
Marks, L.U. (2002), *Touch: Sensuous Theory and Multisensory Media*, University of Minnesota Press, Minneapolis.
Menarini, R. (2015), *Il corpo nel cinema. Storie, simboli e immaginari*, Bruno Mondadori, Milan.
Merleau-Ponty, M. (1945), *Phenomenology of Perception*, trans. Colin Firth, Routledge & Kegan Paul, London.
Merleau-Ponty, M. (2011), *Le monde sensible et le monde de l'expression. Cours au Collège de France. Notes*, 1953, text established and annotated by E. de Saint-Aubert and S. Kristensen, MetisPresses, Geneva.
Metz, C. (1968), *Film Language: A Semiotics of the Cinema*, University of Chicago Press, Chicago (1974).
Metz, C. (1991), *Impersonal Enunciation, or the Place of Film*, Columbia University Press, New York.
Meyer, K., J.T. Kaplan, R. Essex, H. Damasio, A.R. Damasio (2011), "Seeing touch is correlated with content specific activity in primary somatosensory cortex", *Cerebral Cortex*, 21, pp. 2113-2121.
Michaud, P.A. (2000), *Aby Warburg and the Image in Motion*, Zone Books, New York.
Montani, P. (2010), *L'immaginazione intermediale. Perlustrare, rifigurare, testimoniare il mondo visibile*, Laterza, Rome-Bari.
Morel, J.P. (2010), "Le Docteur Toulouse ou le cinéma vu par un psycho-physiologiste (1912-1928)", *1895*, 60, pp. 122-155.
Morin, E. (1956), *The Cinema, or the Imaginary Man*, University of Minnesota Press, Minneapolis, 2005.
Müller, S. (2014), "Embodied cognition and camera mobility in F.W. Murnau's *The Last Laugh* and Fritz Lang's *M*", *Paragraph*, 1, pp. 32-46.
Münsterberg, H. (1915), "Why we go to the movies", *Cosmopolitan*, 60, 22-32.
Münsterberg, H. (1916), *The Photoplay. A Psychological Study*, Dover Publications, Mineola, New York.

Murata, A., L. Fadiga, L. Fogassi, V. Gallese, V. Raos, G. Rizzolatti (1997), "Object representation in the ventral premotor cortex (area F5) of the monkey", *Journal of Neurophysiology*, 78, pp. 2226-2230.

Murata, A., V. Gallese, G. Luppino, M. Kaseda, H. Sakata, H. (2000), "Selectivity for the shape, size and orientation of objects in the hand-manipulation-related neurons in the anterior intraparietal (AIP) area of the macaque", *Journal of Neurophysiology*, 83, pp. 2580-2601.

Murch, W. (1995), *In the Blink of an Eye: A Perspective on Film Editing*, Silman-James Press, Hollywood, 2001.

Naci, L., R. Cusack, M. Anello, A.M. Owen (2014), "A common neural code for similar conscious experience in different individuals", *Proceedings of the National Academy of Science of the U S A*, 39, pp. 14277-14282.

Nagel, T. (1974), "*What is it like to be a bat?*" The Philosophical Review, 83, pp. 435-450.

Nakano, T., S. Kitazawa (2010), "Eyeblink entrainment at breakpoints of speech", *Experimental Brain Research*, 205, pp. 577-581.

Nakano, T., Y. Yamamoto, K. Kitajo, T. Takahashi, S. Kitazawa (2009), "Synchronization of spontaneous eyeblinks while viewing video stories", *Proceedings of the Royal Society B Biological Sciences*, 276, pp. 3635-3644.

Niedeggen M., P. Wichmann, P. Stoerig (2001), "Change blindness and time to consciousness", *European Journal of Neuroscience*, 14, pp. 1719-1726.

Nielsen, J.I. (2007), "*Camera Movement in Narrative Cinema: Toward a Taxonomy of Functions*", Ph.D. Thesis, Department of Information and Media Studies, Faculty of Arts, University of Aarhus.

Nieuwenhuis, S., R. Ridderinkhof, J. Blom, G. Band, A. Kok (2001), "Error-related brain potentials are differentially related to awareness of response errors: Evidence from an antisaccade task", *Psychophysiology*, 38, pp. 752-760.

Nishimoto, S., A.T. Vu, T. Naselaris, Y. Benjamini, B. Yu, J.L. Gallant (2011), "Reconstructing visual experiences from brain activity evoked by natural movies", *Current Biology*, 19, pp. 1641-1646

Norden, E. (1968), "Interview with Stanley Kubrick", *Playboy*, 9.

North, D. (2008), *Performing Illusions: Cinema, Special Effects and the Virtual Actor*, Wallflower, London.

Odin, R. (2010), "È giunta l'era del linguaggio cinematografico", *Bianco & Nero*, 568, pp. 7-17.

Oh, J., S.-Y. Jeong, J. Jeong (2012), "The timing and temporal patterns of eye blinking are dynamically modulated by attention", *Human Movement Science*, 31, pp. 1353-1365.

Onians, J. (2007), "Neuroarchaeology and the origins of representation in the Grotte de Chauvet", in C. Renfrew, I. Morley (eds.), *Image and Imagination: A Global Prehistory of Figurative Representation*, MacDonald Institute for Archaeological Research, Cambridge, pp. 307-320.

Ortiz, M.J. (2014), "Visual manifestation of primary metaphors through *mise-en-scène* techniques", *Image & Narrative*, 1, pp. 5-16.

Ortiz, M.J., J. Moya (2015), "The action cams phenomenon: A new trend in audiovisual production", *Communication & Society*, 28, pp. 51-65.

Ortoleva, P. (2013), "Una specie di transfert. Il piacere dello spettacolo e le sue basi ludiche", in R. Fanciullacci, C. Vigna (eds.), *La vita spettacolare. Questioni di etica*, Orthotes Editrice, Napoli-Salerno, pp. 39-57.

Panksepp, J. (1998), *Affective Neuroscience: The Foundation of Human and Animal Emotions*, Oxford University Press, Oxford.

Panksepp, J., L. Biven, (2012), *The Archaeology of Mind. Neuroevolutionary Origins of Human Emotions*, W.W. Norton Publishers, New York.
Panofsky, E., W.S. Heckscher, (1996), *Three Essays on Style*, The MIT Press, Cambridge, MA.
Parsons, L., P.T. Fox, J.H. Downs, T. Glass, T.B. Hirsch, C.C. Martin, P.A. Jerabek, J.L. Lancaster (1995), "Use of implicit motor imagery for visual shape discrimination as revealed by PET", *Nature*, 375, pp. 54–58.
Pearlman, K. (2009), *Cutting Rhythms: Shaping the Film Edit*, Focal Press, Burlington.
Perry, A., S. Bentin (2009), "Mirror activity in the human brain while observing hand movements: A comparison between EEG desynchronization in the mu-range and previous fMRI results", *Brain Research*, 1282, pp. 126–132.
Perry, A., N.F. Troje, S. Bentin (2010), "Exploring motor system contributions to the perception of social information: Evidence from EEG activity in the mu/alpha frequency range", *Social Neuroscience*, 5, pp. 272–284.
Pescatore, G. (2001), *Il narrativo e il sensibile. Semiotica e teoria del cinema*, Hybris, Bologna.
Pfurtscheller, G., F.H. Lopes da Silva (1999), "Event-related desynchronization and related oscillatory phenomena of the brain", in *Handbook of Electroencephalography and Clinical Neurophysiology*, Vol. 6, Elsevier, Amsterdam.
Phillips, H.A. (1914), *The Photodrama*, The Stanhope-Dodge Publishing Company, Larchmont.
Piccino, C. (2009), "Un cellulare contro il 'senso comune'", *il manifesto*, 11 August.
Pinker, S. (1994), *The Language Instinct. How the Mind Creates Language*, William Morrow and Company, New York.
Pinker, S. (1997), *How the Mind Works*, W.W. Norton & Co., New York.
Pinotti, A. (2011), *Empatia. Storia di un'idea da Platone al postumano*, Laterza, Rome-Bari.
Pisters, P. (2012), *The Neuro-Image: A Deleuzian Filmphilosophy of Digital Screen Culture*, Stanford University Press, Stanford.
Pitcher, D., L. Garrido, V. Walsh, B.C. Duchaine (2008), "Transcranial magnetic stimulation disrupts the perception and embodiment of facial expression", *Journal of Neuroscience*, 28, pp. 8929–8933.
Plantinga, C. (1999), "The scene of empathy and the human face on film", in C. Plantinga, G.M. Smith (eds.), *Passionate Views: Film, Cognition, and Emotion*, The John Hopkins University Press, Baltimore, pp. 237–255.
Plantinga, C. (2015), "Facing others: Close-ups of faces in narrative films and in *The Silence of the Lambs*", in L. Zunshine (ed.), *The Oxford Handbook of Cognitive Literary Studies*, Oxford University Press, New York, pp. 291–312.
Polan, D. (2007), *Scenes of Instructions: The Beginnings of US Study of Film*, University of California Press, Berkeley and Los Angeles.
Porro, C. A., M.P. Francescato, V. Cettolo, M.E. Diamond, P. Baraldi, C. Zuiani, M. Bazzocchi, P.E. di Prampero (1996), "Primary motor and sensory cortex activation during motor performance and motor imagery. A functional magnetic resonance study", *Journal of Neuroscience* 16, pp. 7688–7698.
Pravadelli, V. (2007), *La grande Hollywood. Stili di vita e di regia nel cinema classico americano*, Marsilio, Venice.
Pulvermüller, F., L. Fadiga (2010), "Active perception: Sensorimotor circuits as a cortical basis for language", *Nature Reviews Neuroscience*, 11, pp. 351–360.
Rancière, J. (2011), *Mute Speech: Literature, Critical Theory, and Politics*, Columbia University Press, New York.

Raos V., M.A. Umiltà, L. Fogassi, V. Gallese (2006), "Functional properties of grasping-related neurons in the ventral premotor area F5 of the macaque monkey", *Journal of Neurophysiology*, 95, pp. 709-729.

Rainey, J. (2015), "Pixar's John Lasseter says iPhone and GoPro could be next film breakthroughs", *Variety*, May 12.

Raz, G., T. Hendler (2014), "Forking cinematic paths to the self: Neurocinematically informed model of empathy in motion pictures", *Projections*, 2, pp. 89-114.

Raz, G., Y. Jakob, T. Gonen, Y. Winetraub, E. Soreq, T. Flash, T. Hendler (2014), "Cry for her or cry with her: Context-dependent dissociation of two modes of cinematic empathy reflected in network cohesion dynamics", *Social Cognitive and Affective Neuroscience*, 9, 30-38.

Reed, C.L., R.J. Caselli (1994), "The nature of tactile agnosia: A case study", *Neuropsychologia*, 32, pp. 527-539.

Reed, C.L., R.J. Caselli (1995), "Tactile agnosia", *Brain*, 119, pp. 875-888.

Reeves, B., A. Lang, E.Y. Kim, D. Tatar (1999), "The effects of screen size and message content on attention and arousal", *Media Psychology*, 1, pp. 49-67.

Rizzolatti G., G. Luppino (2001), "The cortical motor system", *Neuron*, 31, pp. 889-901.

Rizzolatti G., C. Sinigaglia (2006), *Mirrors in the Brain*, Oxford University Press, Oxford.

Rizzolatti, G., C. Sinigaglia (2010), "The functional role of the parieto-frontal mirror circuit: Interpretations and misinterpretations", *Nature Review Neuroscience*, 11, pp. 264-274.

Rizzolatti G., L. Fogassi, V. Gallese (2001), "Neurophysiological mechanisms underlying the understanding and imitation of action", *Nature Neuroscience Reviews*, 2, pp. 661-670.

Rizzolatti, G., L. Fadiga, V. Gallese, L. Fogassi (1996), "Premotor cortex and the recognition of motor actions", *Cognitive Brain Research*, 3, pp. 131-141.

Rizzolatti G., L. Fadiga, L. Fogassi, V. Gallese (1997), "The space around us", *Science*, 277, pp. 190-191.

Rizzolatti, G., L. Fogassi, V. Gallese (2004), "Cortical mechanisms subserving object grasping, action understanding and imitation", in M.S. Gazzaniga (ed.), *The New Cognitive Neurosciences*, 3rd edn, The MIT Press, Cambridge, pp. 427-440.

Rochat, M.J., F. Caruana, A. Jezzini, L. Escola, I. Intskirveli, F. Grammont, V. Gallese, G. Rizzolatti, M.A. Umiltà (2010), "Responses of mirror neurons in area F5 to hand and tool grasping observation", *Experimental Brain Research*, 204, pp. 605-616.

Rodowick, D.N. (2007), *The Virtual Life of Film*, Harvard University Press, Cambridge.

Rodrigo, P. (2013), "Voir et toucher. L'optique, l'haptique et le visuel chez Merleau-Ponty", in M. Carbone (under the direction of), *L'empreinte du visuel. Merleau-Ponty et les images aujourd'hui*, MetisPresses, Geneva, pp. 27-41.

Roland, P., B. Larsen, N. Lassen, E. Skinhoj (1980), "Supplementary motor area and other cortical areas in organization of voluntary movements in man." *Journal of Neurophysiology* 43, pp. 118-136.

Rose, F. (2011), *The Art of Immersion. How the Digital Generation is Remaking Hollywood*, W.W. Norton & Company, New York.

Rose, F. (2013), "Movies of the future", *The New York Times*, June 22, 2013.

Rothstein, P., R.N. Henson, A. Treves, J. Driver, R.J. Dolan (2005), "Morphing Marilyn into Maggie dissociates physical and identity face representations in the brain", *Nature Neuroscience*, 1, pp. 107-113.

Ruchsow M., M. Spitzer, G. Grön, J. Grothe, M. Kiefer (2005), "Error processing and impulsiveness in normals: evidence from event-related potentials", *Brain Research Cognitive Brain Research*, 24, pp. 317-325.

Salt, B. (1993), *Film Style & Technology: History and Analysis*, Starword, London.
Sargent, E.W. (1913), *The Technique of the Photoplay*, The Moving Picture World, New York.
Sarris, A. (1967), *Interviews with Film Directors*, Avon Books, New York.
Schaefer, M., B. Xu, H. Flor, L.G. Cohen (2009), "Effects of Different Viewing Perspectives on Somatosensory Activations During Observation of Touch", *Human Brain Mapping*, 30, pp. 2722-2730.
Schaefer, M., H.J. Heinze, M. Rotte (2012), "Embodied empathy for tactile events: Interindividual differences and vicarious somatosensory responses during touch observations", *Neuroimage*, 60, 952-957.
Schefer, J.L. (1980), *The Ordinary Man of Cinema*, Semiotext(e), Los Angeles, 2016.
Schnitzler, A., S. Salenius, R. Salmelin, V. Jousmäki, R. Hari (1997), "Involvement of primary motor cortex in motor imagery: A neuromagnetic study." *NeuroImage* 6, pp. 201-208.
Schwan S., S. Ildirar (2010), "Watching film for the first time: How adult viewers interpret perceptual discontinuities in film", *Psychological Science*, 21, pp. 970-976.
Seesslen, G. (2004), "Shoot me. Shoot me. On the essence of war: *Full Metal Jacket*", in H.P. Reichmann (ed.), *Stanley Kubrick*, Deutsches Filmmuseum, Frankfurt am Main, pp. 208-223.
Sestito, M., M.A. Umiltà, G. De Paola, R. Fortunati, A. Raballo, E. Leuci (2013), "Facial reactions in response to dynamic emotional stimuli in different modalities in patients suffering from schizophrenia: A behavioral and EMG study", *Frontiers in Human Neuroscience*, 7, pp. 1-12.
Severi, C. (2004), *Il percorso e la voce. Un'antropologia della memoria*, Einaudi, Turin.
Shaviro, S. (1993), *The Cinematic Body*, University of Minnesota Press, Minneapolis.
Shepherd, S.V., J.T. Klein, R.O. Deaner, M.L. Platt (2009), "Mirroring of attention by neurons in macaque parietal cortex", *Proceedings of The National Academy of Sciences U S A*, 106, pp. 9489-9494.
Shimamura, A.P. (2013a), "Presenting and analyzing movie stimuli for psychocinematics research", *Tutorials in Quantitative Methods for Psychology*, 1, pp. 1-5.
Shimamura, A.P. (ed.) (2013b), *Psychocinematics: Exploring Cognition at the Movies*, New York, Oxford University Press.
Shimamura, A.P., D.E. Marian, A.L. Haskins (2013), "Neural correlates of emotional regulation while viewing films", *Brain Imaging and Behavior*, 7, pp. 77-84.
Siegle G.J, N. Ichikawa, S. Steinhauer (2008), "Blink before and after you think: Blinks occur prior to and following cognitive load indexed by pupillary responses", *Psychophysiology*, 45, pp. 679-687.
Silvera, D.H., R.A. Josephs, R.B. Giesler (2002), "Bigger is better: The influence of physical size on aesthetic preference judgments", *Journal of Behavioral Decision Making*, 15, pp. 189-202.
Singer, T., B. Seymour, J. O'Doherty, H. Kaube, R.J. Dolan, C.D. Frith (2004), "Empathy for pain involves the affective but not sensory components of pain", *Science*, 303, pp. 1157-1162.
Slevin, J. (1912), *On Picture-Play Writing: a Hand-Book of Workmanship*, Farmer and Smith, Cedar Grove.
Smith, M. (1995), *Engaging Characters: Fiction, Emotion, and the Cinema*, Clarendon Press, Oxford.
Smith, M. (2009), "Consciousness", in P. Livingstone, C. Plantinga (eds.), *The Routledge Companion to Philosophy and Film*, Routledge, New York, pp. 39-51.
Smith, M. (2012), "Triangulating aesthetic experience", in A.P. Shimamura, S.E. Palmer (eds.), *Aesthetic Science: Connecting Mind, Brains, and Experience*, Oxford University Press, New York, 2012, pp. 80-106.

Smith, T.J. (2012), "The attentional theory of cinematic continuity", *Projections*, 1, pp. 1–27.
Smith, T.J., J.M. Henderson (2008), "Edit blindness: The relationship between attention and global change blindness in dynamic scenes", *Journal of Eye Movement Research*, 2, pp. 1–17.
Smith, T.J., D. Levin, J.E. Cutting (2012), "A window on reality: Perceiving edited moving images", *Current Directions in Psychological Science*, 21, pp. 107–113.
Sobchack, V. (1982), "Toward inhabited space: The semiotic structure of camera movement in the cinema", *Semiotica*, 1, pp. 317–335.
Sobchack, V. (1991), *The Address of the Eye: A Phenomenology of Film Experience*, Princeton University Press, Princeton.
Sobchack, V. (2005), *Carnal Thuoghts: Embodiment and Moving Image Culture*, University of California Press, Berkeley and Los Angeles.
Sobchack, V. (2011), "The man who wasn't there: The production of subjectivity in Delmer Daves' *Dark Passage*", in D. Chateau (ed.), *Subjectivity: Filmic Representation and the Spectator's Experience*, Amsterdam University Press, Amsterdam, pp. 69–83.
Solms M., J. Panksepp (2012), "The 'Id' knows more than the 'Ego' admits: Neuropsychoanalytic and primal consciousness perspectives on the interface between affective and cognitive neuroscience", *Brain Science*, 2, pp. 147–175.
Somaini, A. (2011), *Ejzenštejn. Il cinema, le arti, il montaggio*, Einaudi, Turin.
Sontag, S. (1969), *Styles of Radical Will*, Farrar, Strauss, and Giroux, New York.
Stadler, J. (2008), *Pulling Focus: Intersubjective Experience, Narrative Film, and Ethics*, Continuum, New York.
Stancak, A., G. Pfurtscheller (1996), "Mu-rhythm changes in brisk an slow self-paced finger movements", *NeuroReport*, 7, pp. 1161–1164.
Steinhauer, K., J.E. Drury (2012), "On the early left-anterior negativity (ELAN) in syntax studies", *Brain and Language*, 120, pp. 135–162.
Stern, D.N. (1985), *The Interpersonal World of the Infant*, H. Karnac (Books), London.
Stern, D.N. (2010), *Forms of Vitality. Exploring Dynamic Experience in Psychology, the Arts, Psychotherapy and Development*, Oxford University Press, Oxford.
Svankmajer, J., E. Svankmajer (1998), *Anima Animus Animation. Between Film and Free Expression*, Slovart Arbor Vitae Foundation, Prague.
Sweetser, E.E. (1990), *From Etymology to Pragmatics: Metaphorical and Cultural Aspects of Semantic Structure*, Cambridge University Press, Cambridge.
Tan, E.S. (1996), *Emotion and the Structure of Narrative Film: Film as an Emotion Machine*, Routledge, New York.
Taylor, J. (1914), *The Photoplay*, Washington Printing, Washington.
Tedesco, S. (2008), *Forme viventi. Antropologia ed estetica dell'espressione*, Mimesis, Milan-Udine.
Tirard, L. (ed) (2002), *Moviemakers' Master Class. Private Lessons from the World's Foremost Directors*, Farrar, Strauss and Giraux, New York.
Toro, C., G. Deuschl, R. Thatcher, S. Sato, C. Kufta, M. Hallett (1994), "Event related desynchronization and movement-related cortical potentials on the ECoG and EEG", *Electroencephalography and Clinical Neurophysiology*, 93, pp. 380–389.
Toulouse, E., R. Mourgue (1920), "Des réactions respiratoires au cours de projections cinématographique", *1895*, 60, 2010.
Trasatti, S. (1993), *Ingmar Bergman*, Il Castoro, Milan.
Troscianko, T., T.S. Meese, S. Hinde (2012), "Perception while watching movies: Effects of physical screen size and scene type", *i-Perception*, 3, pp. 414–425.
Truffaut, F. (1993), *Hitchcock/Truffaut*, Editions Robert Laffont, Paris.
Tsakiris, M. (2010), "My body in the brain: A neurocognitive model of body-ownership", *Neuropsychologia*, 48, pp. 703–712.

Turner, M. (1996), *The Literary Mind: the Origins of Thought and Language*, Oxford University Press, New York.
Uithol, S., V. Gallese (2015), "The role of the body in social cognition", *Wires Cognitive Science*, 1.
Uithol S., M. Franca, K. Heimann, D. Marzoli, P. Capotosto, L. Tommasi, V. Gallese (2015), "Single-pulse transcranial magnetic stimulation reveals contribution of premotor cortex to object shape recognition", *Brain Stimulation*, 8, 953-956.
Umiltà, M.A., E. Kohler, V. Gallese (2001), "I know what you are doing: A neurophysiological study", *Neuron*, 32, pp. 91-101.
Umiltà, M.A., T. Brochier, R.L. Spinks, R.N. Lemon (2007), "Simultaneous recording of macaque premotor and primary motor cortex neuronal populations reveals different functional contributions to visuomotor grasp", *Journal of Neurophysiology*, 98, pp. 488-501.
Umiltà, M.A., L. Escola, I. Intskirveli, F. Grammont, M. Rochat, F. Caruana, A. Jezzini, A., V. Gallese, G. Rizzolatti (2008), "How pliers become fingers in the monkey motor system", *PNAS*, 105, pp. 2209-2213.
Valenza, N., R. Ptak, I. Zimine, M. Badan, L. Lazeyras, A. Schnider (2001), "Dissociated active and passive tactile shape recognition: a case study of pure tactile apraxia", *Brain*, 124, pp. 2287-2298.
Varela, F.J., J. Shear (1999), "The view from within: first-person approaches to the study of consciousness", *Journal of Consciousness Studies*, 6, pp. 2-3.
Vico, G.B. (1725/1744), *La Scienza Nuova*, Rizzoli, Milan.
Voss, C. (2011), "Film experience and the formation of illusion: The spectator as 'surrogate body' for the cinema", *Cinema Journal*, 50, pp. 136-150.
Walker, A. (1971/1999), *Stanley Kubrick, Director*, W.W. Norton & Company, New York.
Walla-Romana, C. (2012), "Epstein's *Photogénie* as corporeal vision: Inner sensation, queer embodiment, and ethics", in S. Keller, J.N. Paul (eds.), *Jean Epstein. Critical Essays and New Translations*, Amsterdam University Press, Amsterdam, pp. 51-71.
Wicker, B., C. Keysers, J. Plailly, J.P. Royet, V. Gallese, G. Rizzolatti (2003), "Both of us disgusted in my insula—the common neural basis of seeing and feeling disgust" *Neuron*, 40, pp. 655-664.
Wilson, G. (2006), "Transparency and film in narrative fiction film", *The Journal of Aesthetics and Art Criticism*, 1, pp. 81-95.
Wojciehowski H.C., V. Gallese (2011), "How stories make us feel. Toward an embodied narratology", *California Italian Studies*, 2, 1.
Zacks, J.M. (2015), *Flicker: Your Brain on Movies*, Oxford University Press.
Zacks, J.M., Magliano, J.P. (2011), "Film, narrative, and cognitive neuroscience", in F. Bacci, D. Melcher (eds.), *Art and the Senses*, Oxford University Press, New York.
Zacks, J.M., N.K. Speer, J.R. Reynolds (2009), "Segmentation in reading and film comprehension", *Journal of Experimental Psychology: General*, 138, pp. 307-327.
Zaki, J., J. Weber, N. Bolger, K. Ochsner (2009), "The neural bases of empathic accuracy", *Proceedings of the National Academy of Sciences USA*, 106, pp. 11382-11387.
Zeki S. (1999), "Art and the brain", *Journal of Consciousness Studies*, 6, pp. 76-96.
Zeki, S. (2009), *Splendours and Miseries of the Brain. Love, Creativity and the Quest for Human Happiness*, John Wiley and Sons, Hoboken.
Zernik, C. (2012), *L'œil et l'objectif. La psychologie de la perception à l'épreuve du style cinématographique*, Vrin, Paris.
Zhou, J.M., Y.D. Fuster (2000), "Visuo-tactile cross-modal associations in cortical somatosensory cells", *Proceedings of the National Academy of Sciences of the United States of America*, 97, pp. 9777-9782.

Index

Figures are indicated by an italic *f* following the page number.
Foreign works are alphabetized under English titles.
For the benefit of digital users, indexed terms that span two pages (e.g., 52–53) may, on occasion, appear on only one of those pages.

Numbers
180-degree rule 134–36, 208
   investigation of 136
      ERD results 142–44
      ERP results 138–41
      experimental method 137–38

action 126
   chain of 128
   facilitation of narration 129–30
   importance in screenplay writing 130
   line of 131–32
   meanings of 126–28, 130
   psychological action 131
   role in intersubjectivity 126, 130–31
   syntactic violations of structure 138–39
   see also continuity editing
action cameras (GoPros) 203, 194–95, 198–99
action films 130–31
affective neuroscience 4n7
affordance 25–27n36
Alcott, John 101
alien movements, embodied simulation of 201–2
alignment 66–68
Allard, Laurence 196
Allegri, Luigi 166n53
Amengual, Barthélemy 59–60
Anderson, Joseph 45–46, 121–22
Andersson, Bibi 145–47
Andersson, Roy 91–92
animals, empathy with 1
animation 175*f*
   computer-generated imagery 175–78
   Jan Švankmajer's work 174*f*, 174–76
   motion capture 178–79

anterolateral system 154*f*
Antiquity (Marc Azéma and Florent Rivère, 2012) xv–xvi
Aristotle 40–41
Arnheim, Rudolf 94–95, 97n40
L'arrivée d'un train en gare à La Ciotat (Lumière Brothers, 1895) 71
Art of the Moving Picture, The (Linsay Vachel, 1915) 130–31
Artaud, Anonin 73
associative areas 12–13
attention 124–25
Au hasard Balthazar (Robert Bresson, 1966) 1, 2–3
auditory cortex 152*f*
Auerbach, Jonathan 129
Auster, Paul 126
axial cuts 203
Azéma, Marc xv–xvi

Balàzs, Béla 89–90, 148–49, 177–78
Ball, Eustace Hale 128, 132
Baranowski, Andreas 186–87
Barker, Jennifer 145, 177–78
barrier, screen as 200
Barry Lyndon (Stanley Kubrick, 1975) 98–99
Barrymore, Ethel 63
Bay, Michael 91
Bazin, André xviii–xix
Bellman, Steven 185–86
Bellmer, Hans 123n8
Beowulf (Robert Zemeckis, 2007) 207
Bergman, Ingmar 145–48, 164, 165
Bergman, Ingrid 53
Bertolucci, Bernardo 92, 115–16
Bettman, Gil 85, 91, 92–93

Black Mirror (TV series) 197n30
blinking 133–34
   during viewing of a movie 134
Bob's Rules 92–93, 115
body, dual nature of 11
body shots 129
Bogart, Humphrey 60
Book of Illusions, The (Paul Auster, 2002) 126
Bordwell, David 46–47, 54–55, 86, 87*f*, 87–88, 149–50
Botticelli, Sandro 7
Bracken, Cheryl 185–86
brain imaging 203
   see also electroencephalography (EEG) studies; fMRI
brain–body system 11
   and cinema 44
Branagh, Kenneth 93–94
Bredekamp, Horst 190
Brent, George 63
Bresson, Robert 1, 2–3
Brown, Garrett 95, 99–100, 208
Bruno, Marcello Walter 97–98
Buccheri, Vincenzo 88–89
Buccino, Giovanni 201
Bush, George H.W. 46

camera action 131
camera movement 91
   circular 93–94
   in creation of suspense 55–59
   EEG responses
      event-related desynchronization (ERD) 110–11
      experimental method 108–9, 109*f*, 110*f*
      objective 108
      results 111–14, 112*f*
   first-person narrative style 59–61, 198–99
   and fundamental movements of cinema 96–97, 97n40
   invisible 92–93, 96
   Kubrick's films
      Barry Lyndon (1975), use of zoom 98–99
      Full Metal Jacket (1987) 104–5
      The Shining (1980), use of Steadicam 99–104

   metaphoric value 94–95
   motor simulation 58–59
   and operator's bodily movements 95
   role of the camera 92
   static style 91–92
   Steadicam, corporeality of 114–16
Camera-I (first-person camera) 59–61
   see also first-person narrative style
Cameron, James 91
canonical neurons 24, 26*f*, 69–70
Carluccio, Giulia 89
Carroll, Noel 46–47, 57n6
Casebier, Alan 46
Casetti, Francesco 62–63, 88, 150, 183–84
Casino (Martin Scorsese, 1995) 115
cave drawings xv
Cave of Forgotten Dreams (Werner Herzog, 2010) xv
Cinematografo cerebrale (*Cerebral Cinematograph*) (Edmondo De Amicis, 1907) 72–73
characters in action 128n25, 128
chase movies 130–31, 132n39
Chateau, Dominique 116
Chauvet cave drawings xv
Christmas Carol, A (Robert Zemeckis, 2009) 207
Ciment, Michel 101
Cinderella (Kenneth Branagh, 2015) 93–94
The Cinema, or the Imaginary Man (Edgar Morin, 1956) 74–75
cinesthetic subject, the 76–77, 79
circular camera movement 93–94
classical cognitivism 12–13
   sandwich model 13*f*
clay animation 174–76
Clockwork Orange, A (Stanley Kubrick, 1971) 98, 101
close-ups xxiii, 91–92
   Bergman's use of 147–48
   embodied simulation model 150–51
   Godard's use of 166–70
   Švankmajer's use of 172*f*, 172, 173*f*, 175
   theories of 148–50
coenesthesia 74–76
cognitive modules 13–14
   Swiss Army knife metaphor 14*f*
cognitive neuroscience 10
cognitive studies 46–47

INDEX 229

Columbus, Chris 51
comparative perspective 11–12
Complete Action 130
computer-generated imagery (CGI),
　tactility 175–76
　Toy Story (John Lasseter, 1995) 177–78
connections 121–22
continuity 121
　line of action 131–32
　in visual perception 133–34
continuity editing 51, 120–22, 125, 133, 203
　180-degree rule 134–36
　　investigation of 136–44
　exploitation of physiological
　　characteristics 134–36
　The Silence of the Lambs (Jonathan
　　Demme, 1991) 117–20
convergence 182–83
corporeal images 196
corporeality *see* embodiment
corridors, Kubrick's use of 103*f*, 104
Cosmides, Leda 14
Cronenberg, David 115–16
crossing the line 134–36
　EEG investigation of 136–44
　see also jump cuts
cross-modal confirmation 121
cuneate nucleus 153*f*
Currie, Greg 78
Cutting, James 46–47, 132

D'Aloia, Adriano 47, 89–90, 108n55
Damasio, Antonio 45–46, 75–76
"Dance" experiment, Lev Kulešov, 123–24
Dark Passage (Delmer Daves,
　1947) 60–61, 99–100
Tma/Svetlo/Tma (*Darkness Light Darkness*,
　Jan Švankmajer, 1989) 174–76, 176*f*
Darwin, Charles 7
Daves, Delmer 60–61, 99–100
De Amicis, Edmondo 72–73
De Palma, Brian 191–92
Del Bono, Pippo 197
Deleuze, Gilles 45–46, 147n2, 147–48, 177–78
Demme, Jonathan 117–20, 122–23
depth-of-field 204
Devil's Cinema, The (Jean Epstein,
　1947) 76n39

Moznosti dialogu (*Dimensions of dialogue*,
　Jan Švankmajer, 1982) 174*f*,
　174, 175*f*
direction editing 203
disconnection from the world 43
disgust, fMRI studies 34–37
dolly
　distinction from Steadicam 114–15
　EEG responses 112*f*
　for smartphones 196
　spectators' evaluation of 113*f*
dolly shots 207
dorsal column–medial lemniscal
　system 153*f*
Dr Jekyll and Mr Hyde (Rouben
　Mamoulian, 1931) 59
Dreyer, Carl Theodor 91–92
Dudai, Yadin 45–46
duplication hypothesis 73–74
dynamograms 8

Ebisch, Sjoerd 159–63
editing 125
　spatiality and temporality effects 123–24
　studies of 51
　see also continuity editing
Ejzenštejn, Sergej 71n22, 71, 124, 149–50
electroencephalography (EEG)
　studies 15, 107–8, 204
　of 180-degree rule 137
　　experimental method 137–38
　　results 138–41
　of effects of camera movements
　　event-related desynchronization
　　　(ERD) 110–11
　　experimental method 108–9,
　　　109*f*, 110*f*
　　objective 108
　　results 111–14, 112*f*
Elsaesser, Thomas 147–48, 177–78
embodied cognition 16–17
embodied simulation xix–xxi, 1–3, 48–49,
　67–68, 150–51, 181, 204
　action cam videos 199
　of alien movements 201–2
　and the cinesthetic subject 77
　and close-ups 150–51
　emotions and sensations 33
　and imagination 38–41

embodied simulation (*cont.*)
  jump cuts, effect of 136
  liberated xxi, 40–44
  manipulable objects perception 24
  mirror neuron mechanism 27, 28*f*
  and motion capture 179
  peri-personal space 21, 69–70
  and presence 188–89
  and the Steadicam 114–15
  tactile cinema 170
  in *Unfriended* (Levan Gabriadze, 2014) 193–94
embodiment xvii, 70–71, 205
  circular camera movement 93–94
  forms of 127
Emerson, John 128
emotional transfer 43–44
emotions
  corporeal expression of 7
  embodied simulation 37
  facial mimicking 33–34
  fMRI studies 34–37
empathy 5, 145, 170, 205
  in aesthetic enjoyment 5–6
  with animals 1
  fMRI studies 51
  relationship to embodied simulation 9, 39–40
  role of somatosensory cortex 159–60
engram concept 8
enteroception 4–5
epigenetics 7–8
Epstein, Jean 76n39, 124–25, 171–72
establishing shots 207
Eugeni, Ruggero 47, 104, 108n55
event segmentation 51
event-related desynchronization (ERD) 204
  response to camera movements 110–11, 114
  response to editing 142–44
event-related potentials (ERPs) 138, 204
  response to editing 138–41
  response to syntactic violation of action structure 138–39
evolutionistic psychology 13–14
excess practices 62–63
experimental aesthetics approach xviii

Expression of the Emotions in Man and Animals, The (Charles Darwin, 1872) 7
externally generated camera movement 93
extreme long shots (ELS) 207
extreme wide shots (EWS) 207
eye movements 133
  mirror mechanism 31–32
  during viewing of a movie 134
eye trace editing 203

F4 neurons 21–24
  somatosensory and visual receptive fields 22*f*, 23*f*
F5 neurons 19
  canonical neurons 24, 26*f*
  mirror neurons 27, 28*f*
"Fabricated Landscape" experiment, Lev Kulešov 123–24
"Fabricated Person" experiment, Lev Kulešov 123–24
face
  Bergman's depiction of 147n2, 147–48
  see also close-ups
facial expression
  fMRI studies 34–37
  motion-capture technology 178–79, 207
  "Možžuchin" experiment 123–24
facial mimicking 33–34
Zánik domů Usherů (The Fall of the House of Usher, Jan Švankmajer, 1980) 171–72, 172*f*, 173*f*
Farassino, Alberto 165
Feldmann, Erich 73–74
Ferraris, Maurizio 196
figure tensive (tensive forms) 89–90
Film. A Psychological Study (Hugo Münsterberg, 1916) xviii–xix
film cognition, levels of 87*f*
film noir 63–66
filmologie 47
firewall hypothesis 54–55
first-person narrative style 59–61
  action cams 198–99
Flusser, Vilém 182
fMRI (functional magnetic resonance imaging) 206

fMRI studies 14–15
  on editing 141
  on empathic involvement 51
  on facial expression 34–37
  on seeing a dog barking 201
  on touch 159–63, 160f, 161f
Fonteyn, Margot 95
Foscolo, Ugo 70
Foster, Jodie 117–20
found-footage horror 190–94
fragmentation 121
Freeburg, Victor 54, 91, 124–25, 128, 130
Freud, Sigmund 40–41
Friedberg, Anne 182–83
Full Metal Jacket (Stanley Kubrick, 1987) 104–5, 106f
full shots (FS) 207
fundamental movements of cinema 96–97, 97n40
Fuster, Joaquin 158

Gabriadze, Levan (Leo) 190–94
Gallese, Vittorio 136n52
Gandini, Leonardo 85n6
Gaudreault, André 129
Gestaltkreis concept 8n13
gestures 7, 58, 129–30, 169–70
  mirror neuron activation 27
Geuens, Jean-Pierre 115
Ghanzanfar, Asif 45–46, 47
Ghezzi, Enrico 98–99, 99n45
Gibson, J.J. 121–22
Glenn, Scott 117–20
goal sensitivity, premotor neurons 19
Godard, Jean-Luc 165–70
Goldman, Alvin 9n14
Goldstein, Kurt 170–71
The Good, the Bad and the Ugly (Sergio Leone, 1966) 186–87
Goodfellas (Martin Scorsese, 1990) 115
Goodman, Nelson 85
Google Glass 196–97
GoPro cameras (action cameras) 194–95, 198–99, 203
Gorkij, Maksim 72–73, 82
Gottschall, Jonathan 186–87
gracile nucleus 153f
Grand Theory 46
Grant, Cary 53

"grasp on the world" 79
Griffith, David Wark 92
Grodal, Torben 78, 148–49
Gumbrecht, Hans 188–89

Hagener, Malte 147–48, 177–78
hand movements 170–71
  exploratory 158–59
  haptic visuality 165–70
  mirror neuron responses 28f, 30f, 201
  smartphone manipulations 190
  see also gestures
hand-held camera, Kubrick's use of 98, 101, 104–5
hands 145
  in first-person camera technique 60–61
  responses to touch of 159–61, 160f
Haneke, Michael 2–3
Hannon, William Morgan 131–32
Hansen, Mark 189
haptic era, new mediation 188–90
haptic vision 150, 188–89
haptic visuality 164
  Jan Švankmajer's work 170–76
  Une femme mariée (*A Married Woman*, Jean-Luc Godard, 1964) 166–70
  motion capture 178–79
Hasson, Uri 46–47, 50–51, 186–87
Hecht, Heiko 186–87
Heidegger, Martin 25–27, 69, 196
Heimann, Katrin 136n52
Hemingway, Anthony 198
Hennig, Shelley 191–92
Hering, Ewald 7–8
Herzog, Werner xv
high-density EEG
  see electroencephalography (EEG) studies
Hildebrand, Adolf von 6
Hiroshima mon amour (Alain Resnais, 1959) 165
Histoire(s) du cinéma (Jean-Luc Godard, 1980) 165
Hitchcock, Albert 53–59, 92
Højbjerg, Lennard 93–94
Hurley, Susan 12
Husserl, Edmund 76–77, 157, 158
Huston, John 85
Hutchison, William 37n56

Ihde, Don 71
imagination 38–41
  in non-human species 40–41
imaging *see* electroencephalography (EEG) studies; fMRI
immersive properties of cinema 73–74
  new mediation 200, 201n38, 201–2
immobility, intensification of simulation 43–44
imprint (Prägung) concept 7
incredulity 42n64
individual neuron recordings 16
Ingarden, Roman 76–77
institutional mode of representation (IMR) 121–22
insula 34–37
  response to observation of touching 160–61
intensive stylistics 88–89
intentional consonance 4–5
intercorporeality 3–4
internally generated camera movement 93, 103–4
interpersonal relationships 4–5
intersubjectivity xvi, 1–2, 5, 10, 17, 44, 49–50, 54–55, 61–62, 204, 206
  in animation 167f, 177–79
  mirror neurons 3–4, 29, 33
  role of action 126, 130–31
iPods
  video viewing 185–86
  see also new mediation

Jackson, Peter 91, 178–79, 207
Jeanerod, Marc 170–71, 204
Jenkins, Henry 194–95
Johnson, Mark 48, 70, 94
Jonas, Hans 158–59
jump cuts 135f, 136, 206, 208
  event-related desynchronization 142–44
  event-related potentials 141

Kant, Immanuel 6
Keaton, Buster 62–63, 80, 81f, 200
King Kong (Peter Jackson, 2005) 178–79, 207
Koch, Christof 75
Körper 11

Kracauer, Siegfried 76–77, 77n43
Kubrick, Stanley 92, 97
  Barry Lyndon (1975) 98–99
  Full Metal Jacket (1987) 104–5
  The Shining (1980) 99–104
Kulešov, Lev 123–24
Kulešov Effect 123n7, 123–24, 124n9

La Paura (Pippo Del Bono, 2009) 197
Lady in the Lake (Robert Montgomery, 1947) 59f, 59
Lageira, Jacinto 73–74
Lakoff, George 94
Lang, Fritz 63–64
language 40–41
laptops 182
  watching a movie on 184–85, 192–94
  see also new mediation
Lasseter, John 177, 194–95
Leib 11, 157, 158
Leonardo da Vinci 190
Leone, Sergio 186–87
Leroy, Philippe 165
lesion studies 16
Levine, Ted 117–20
liberated simulation xxi, 40–44
Lindsay, Vachel 130–31
line of action 131–32
linguistic metaphors, fMRI study 163
linguistic representation 9
Lipps, Theodore 5
Lo Piparo, Franco 40–41
location of viewing 186–87
long shots 207
  moving 93
  The Shining (Stanley Kubrick, 1980) 103f
Loos, Anita 128
Lord of the Rings film series, (Peter Jackson) 178–79, 207
Lucas, George 198, 199–200

MacDougal, David 196
Mach, Ernst 61, 62f
Magliano, Joseph 46–47, 51, 141
magnetoencephalography (MEG) 15
Mamoulian, Rouben 59
manipulable objects, perception of 24
Marey, Etienne-Jules 8

Marks, Laura 164, 177–78
Une femme mariée (*A Married Woman*,
  Jean-Luc Godard, 1964) 165–70
McConkey, Larry 101
McGinn, Colin 75
McGuire, Dorothy 63
McKay, Craig 122–23
medial lemniscus 153*f*, 154*f*
mediation xxii
medium long shots (MLS) 207
memory 7–8
mental action 131
Méril, Macha 165, 169
Merleau-Ponty, Maurice 24, 46, 76–77,
  78–79, 95–96, 157–58, 170–71
Metz, Christian 73n27, 85, 164, 175–76
Michaud, Philippe Alain 129
Michotte, Albert 76–77
mimesis 131
mind-movie problem 75–76
mirror mechanism 31*f*, 55, 206
  eye movements 31–32
  facial mimicking 33–34
  mu rhythm 110–11, 142
  pain perception 37n56
  plasticity of 201
  response to incompletely seen
    actions 29–31, 30*f*, 40–41
  response to sound 32
mirror neurons 3–4, 27, 28*f*, 106,
  187n17, 206
mirror rule 45
mneme concept 8
mobile viewing 182–87
  see also new mediation
Montani, Pietro 191–92
Montgomery, Robert 59
Morgan, William 129
Morin, Edgar 43, 74–75
motion capture (performance
  capture) 178–79, 207
motor cognition 4
  area F4 and peri-personal space 21
  canonical neurons 24
  mirror neurons 27, 28*f*
motor cortex 17–18, 152*f*
motor intentionality 33
motor possibilities 78–79
motor reactions 71

motor resonance 55, 69, 77n43
  conflicting views on 78
  respiratory reactions 77
motor simulation
  first-person narrative style 59–61
  Notorious (Albert Hitchcock,
    1946) 53–59
  Spiral Staircase, The (Robert Siodmak,
    1946) 63–66
motor system, functions 19, 20*f*
Mourgue, Raoul 77
movement editing 203
movies
  essential components 44–45
  mobile viewing 182–87
  neuroscientific approaches 46–48
  power of 49
  reactions to 45–46
  reproduction of reality 44
Moya, José A. 198
"Možžuchin" experiment, Lev
  Kulešov 123–24
mu rhythm 110–11
  response to editing 142–44
multimodal neural integration 156–57, 158
multimodality 76–77
multiplicity of cinema 49
Münsterberg, Hugo xviii–xix, 124–25, 131,
  132n42, 148–49
Murch, Walter 121, 124–25
Murnau, Friedrich 92

Nachleben concept 7–8
Nagel, Thomas 200–1
natural action 128
negotiation 88
neoteny 43n65
neuro-aesthetics xvii
neurocinematics 47, 48, 50–51
neurofilmology 47, 48
neuronal recycling 1–2, 2n1
neurons 207
  see also mirror neurons; premotor neurons
neuroscience 10
  use of video stimuli 50–51
neuroscientific approaches xvii–xxi, 10
  brain-body system 11
  classical cognitivism 12–13
  comparative perspective 11–12

neuroscientific approaches (*cont.*)
  embodied cognition 16–17
  evolutionistic psychology 13–14
  experimental methodologies 15
  fMRI 14–15
  individual neuron recordings 16
  lesion studies 16
  non-human primate studies 16
neuroscientific studies
  purpose of 107–8
  see also electroencephalography (EEG) studies; fMRI studies
new mediation 182
  adaptation to 186–87
  corporeal interactions with 187
  GoPro cameras 198–99
  immersive potential 187
  implications for movie industry 199
  location of viewing 186–87
  screen size 185–86
  smartphones 194–98
  Unfriended (Levan (Leo) Gabriadze, 2014) 190
Paranmanjang, (*Night Fishing*, Park brothers, 2012) 197
nociception, anterolateral system 154*f*
Noël, Bernard 165
Nolan, Chris 91
non-human primate studies 16
North, Dan 175–76
Notorious (Albert Hitchcock, 1946) 53–59, 69–70, 122–23
Nureyev, Rudolf 95

Oculus Rift 200
off-screen 207
Oliver, Gordon 63
On Picture-Play Writing: A Handbook of Workmanship (James Slevin, 1912) 128n25
Über das Optische Formgefühl: Ein Beitrag zur Aesthetik (*On the Optical Sense of Form: A Contribution to Aesthetics,* Robert Vischer, 1873) 5
Ophüls, Max 92
Ortiz, Maria J. 94n30, 198
over-the shoulder shots (semi-subjective shot) 207

P4-6 potential 139–41
Pabst, Georg Wilhelm 63–64
pain, embodied simulation circuit for 51
pain avoidance 40–41
pain perception
  embodied simulation 37–38
  mirror mechanism 37n56
Pakula, Alan J. 51
Panofsky, Erwin 88–89
parallel sequences 122–23
parietal cortex 18*f*
Park Chan-Kyong 197
Park Chan-Wook 197
Pasolini, Pierpaulo 91–92
Passion of Joan of Arc, The (Carl Theodor Dreyer, 1928) 91–92
Paths of Glory (Stanley Kubrick, 1957) 101
perception 40–41, 124–25, 157–58
  classical cognitivism model 12–13
  embodied simulation 37–38
  of manipulable objects 24
  of peri-personal space 21
  see also somatosensory system; visual perception
peri-personal space 21
  motor resonance 69–70
  new mediation, location within 187
Persona (Ingmar Berman, 1966) 145–48, 164, 165
Pescatore, Guglielmo 149n16
Pettey, Gary 185–86
phantasia sensibile (sensitive phantasy) 40–41
phenomenology 46, 76n42, 90, 164
  application to camera movement 95–96
Phenomenology of Perception (Maurice Merleau-Ponty, 1945) 24, 79, 170–71
Phillips, Henry Albert 130–32
Photodrama, The (Henry Albert Phillips, 1914) 127–28
photogeny 125
Photoplay, The: A Psychological Study (Hugo Münsterberg, 1916) 131
Photoplay Author, The (journal) 125–26
photoplay writing 125–26
  chain of action 128
  importance of action 130
Piccino, Cristina 197

A Pigeon Sat on a Branch Reflecting on Existence (Roy Andersson, 2014) 91–92
Pinker, Steven 14
Pinotti, Andrea 6
pleasure, as stimulus for action 40–41
Poe, Edgar Allan 170–71
point-of-view shots (POV, subjective shots) 66–67, 93, 208
  see also first-person narrative style
Polar Express (Robert Zemeckis, 2004) 207
Porter, Edwin 82
position editing 203
positioning of the spectator 70
  use of new media 182–85
    location of viewing 186–87
    screen size 185–86
Post Theory: Reconstructing Film Studies (David Bordwell & Noel Carroll (eds.), 1996) 46–47
prehistoric "filmmakers" xv–xvi
premotor cortex 152*f*
premotor neurons
  F4 21–24
  F5 19
    canonical neurons 24, 26*f*
    mirror neurons 27, 28*f*
presence 25–27, 38, 73n27, 188–89
  effects of screen size 185–87
  impact of digital media 189, 196–97
primary sensory cortices 152*f*, 156–57
primary somatosensory cortex (SI) 152*f*, 152–54, 153*f*
  response to observation of touching 159–63, 160*f*, 161*f*
  somatosensory homunculus 155*f*
Das Problem der Form in der Bildenden Kunst ("The Problem of Form in Figurative Art", Adolf von Hildebrand, 1893) 6
Project Green Screen (Robert Rodriguez, 2013) 197
projected action 94
psychocinematics 47, 48
psychological action 131

Rains, Claude 53
Rancière, Jacques 49

reality
  reproduction in movies 44
  "virtual" nature of 41
recognition practices 62–63
Red Tails (George Lucas, Anthony Hemingway, 2012) 198
Redacted (Brian De Palma, 2007) 191–92
reductionism 49–50
Reeves, Byron 185–86
relocation of cinema 182–83
  Unfriended (Levan (Leo) Gabriadze, 2014) 192
  see also new mediation
remediation 182–83
Resnais, Alain 165
respiratory reactions 77, 78*f*
reticular formation 154*f*
rhythm of film 124–25
  action as determinant of 128
Rivère, Florent xv–xvi
Rizzolatti, Giacomo 19
Rochat, Magali J. 201
Rodowick, David 186–87
Rodriguez, Robert 197
Rose, Frank 186–87, 200

saccadic eye movements 133
  during viewing of a movie 134
Salt, Barry 87–88, 132
Sargent, Epes Winthrop 127–28, 129, 131, 132
Schefer, Jean-Louis 196–97
Scorsese, Martin 115
screen size 185–86
screenplay writing 125–26
  chain of action 128
  importance of action 130
secondary somatosensory cortex (SII) 153*f*, 156, 158–59
  response to observation of touching 159–60, 161–63
self, sense of 33
semisubjective shots 208
Semon, Richard 7–8
sensations 33, 148–49
  effects of editing 124
  evocation by circular camera movements 93–94
somatosensory system 151

sensations (*cont.*)
    spatial 24
    see also perception; somatosensory system; touch
sensorial impact 72–73
sensuality 73
Serkis, Andrew 178–79, 207
Shepherd, Stephen 31–32
Sherlock Jr. (Buster Keaton, 1924) 62–63, 80, 81*f*, 200
Shimamura, Art 46–47, 50–51, 90
Shining, The (Stanley Kubrick, 1980) 99–104, 102*f*, 103*f*, 115
shot/reverse shot technique 121–22, 208
shots 207–208
SI (primary somatosensory cortex) 152*f*, 152–54, 153*f*
    response to observation of touching 159–63, 160*f*, 161*f*
    somatosensory homunculus 155*f*
SII (secondary somatosensory cortex) 153*f*, 156, 158–59
    response to observation of touching 159–60, 161–63
Tystnaden (*The Silence*) (Ingmar Berman, 1963) 165
Silence of the Lambs, The (Jonathan Demme, 1991) 117–20, 122–23
Siodmak, Robert 63–66
Skin of Film, The (Laura Marks, 2000) 164
'skin of the film' metaphor 164, 177–78
Skin-screen 190n20
Skype 192–93
Slevin, James 128n25, 128
smartphones
    immersive potential 187
    movies shot with 197
    video-photographic properties 194–98
    see also new mediation
Smith, Murray 66–67
Smith, Tim 46–47, 51
Sobchak, Vivian 46, 61, 79, 95–96, 97n40, 114–15, 164
social context xxii
Society for Cognitive Studies of the Moving Image (SCSMI) 46–47
Somaini, Antonio 71, 149–50
somatosensory homunculus 155*f*

somatosensory system 151, 152*f*
    anterolateral system 154*f*
    dorsal column–medial lemniscal system 153*f*
somatosensory homunculus 155*f*
somatotopy 154–56
Sontag, Susan 147
Sophie's Choice (Alan J. Pakula, 1982) 51
Sossaman, Heather 191
sound, mirror neuron responses 32
sound editing 203
Spielberg, Stephen 199–200
spinoreticular tract 154*f*
spinotectal tract 154*f*
spinothalamic tract 154*f*
Spiral Staircase, The (Robert Siodmak, 1946) 63–66, 67
"spirit of the times" 86
Stadler, Jane 169–70
static style 91–92
Steadicam 95, 208
    corporeality 114–16
    Kubrick's use of 99–104
Steadicam movements
    EEG responses 111–14
    spectators' evaluation of 113*f*
Stepmom (Chris Columbus, 1998) 51
Stern, Daniel 33, 149–50
style 85
    bottom-up versus top-down perspective 86–87
    classical American 129
        action as basis of 132
    concept of 85–86
    elements of 90
    and embodied cognition 89–90
    etymology 86
    intensive stylistics 88–89
    social mediation 87–88
    technological mediation 87–88
    and theory of cinema 89
Subjective Perspective (Ernst Mach, 1886) 62*f*
subjective shot 208
subjectivity 3, 4n7, 59–61, 66–68, 74–75, 107n54, 116, 150
    duplication hypothesis 73–74
    first-person narrative style 59–61
    Kubrick's use of 104

of smartphones 196, 197
Unfriended (Levan (Leo) Gabriadze, 2014) 193
success rule 45
surprise 58–59, 66
surrogate intentionality 74
surround hypothesis 186–87
suspense xxiii, 53–54
  continuity editing 117–20, 122–23
  Notorious (Albert Hitchcock, 1946) 53–59
  Spiral Staircase, The (Robert Siodmak, 1946) 63–66
Švankmajer, Jan 170–71
  clay animation 174–76
  Zánik domů Usherů (The Fall of the House of Usher, 1980) 171–72, 172*f*, 173*f*
Sweester, Eve 94
symbiosis, cinema as 74–75
synesthesia 75–76
syntactic violation of action structure 138–39
synthespians 178–79
  see also computer-generated imagery; motion capture (performance capture)

tablets *see* new mediation
tactile agnosia 158–59
tactile cinema 170
tactile exploration 158–59
Tactile Eye, The (Jennifer Barker, 2009) 177–78
tactile sensations
  embodied simulation 37–38
  see also touch
tactile stimuli 152
  dorsal column-medial lemniscal system 153*f*
  see also touch
tactility, computer-generated imagery 175–76
Technique of Photoplay, The (Epes Winthrop Sargent, 1913) 127–28
telematic society 182
television 182–83
Terrone, Enrico 196
thalamus 153*f*
theater, comparison with cinema 130

Theory of Mind, neural correlates 51
thermal sensibility, anterolateral system 154*f*
Tin Toy (John Lasseter, 1988) 177
Tononi, Giulio 75
Tooby, John 14
touch 158
  and digital images 175–76
  haptic visuality 164
  Une femme mariée (A Married Woman, Jean-Luc Godard, 1964) 165–70
  observation of, fMRI studies 159–63, 160*f*, 161*f*
  Švankmajer's work 170–76
  tactile exploration 158–59
  Toy Story (John Lasseter, 1995) 177–78
  in use of new media 186–90
    corporeal interactions 187
Toulouse, Edouard 77
Toy Story (John Lasseter, 1995) 177–78
tracking shots 207
  Kubrick's use of 104–5
transcranial magnetic stimulation (TMS) 15, 208
triangulation 17n26, 51
trigeminal nerve 152
Troscianko, Tom 186–87
Truffaut, François 53–54
Tsivian, Yuri 132n40, 132
Turner, Mark 94

Ullman, Liv 145–47
Umiltà, Maria Alessandra 40–41
Uncle Josh at the Moving Picture Show (Edwin Porter, 1902) 82
Unfriended (Levan (Leo) Gabriadze, 2014) 190–94

Vertov, Dziga 177–78
video stimuli, use in neuroscience 50–51
virtual reality 200
Vischer, Robert 5
visual cortex 152*f*
visual energy 91
visual perception
  intermittency 133–34
  role of embodied simulation 25–27
Vorhandenheit 38–39
Voss, Christiane 74

Walker, Alexander 98–99
Wallon, Henri 76–77
Walton, Kendall 68
Warburg, Aby 6–8
Weizsäcker, Viktor von 8n13
Welles, Orson 38
"Why We Go to the Movies"
    (Hugo Münsterberg,
    1915) 131
Wicker, Bruno 34–36
Wilson, George 68, 120

Zacks, Jeffrey 45, 46–47, 51, 141
Zeki, Semir xvii
Zemeckis, Robert 92–93, 207
Zernik, Clélia 89–90
Zhou, Yong-Di 158
zoom 208
    drawbacks of 115–16
    EEG responses 112*f*
    Kubrick's use of 98–99, 105
    spectators' evaluation of 113*f*
Zuhandenheit 38–39

The manufacturer's authorised representative in the EU for product safety is Oxford University Press España S.A. of el Parque Empresarial San Fernando de Henares, Avenida de Castilla, 2 – 28830 Madrid (www.oup.es/en or product.safety@oup.com). OUP España S.A. also acts as importer into Spain of products made by the manufacturer.

www.ingramcontent.com/pod-product-compliance
Lightning Source LLC
Chambersburg PA
CBHW070647300326
42016CB00036B/118